REJECTION OF EMERGING ORGANIC CONTAMINANTS BY NANOFILTRATION AND REVERSE OSMOSIS MEMBRANES: EFFECTS OF FOULING, MODELLING AND WATER REUSE

T0299981

Rejection of Emerging Organic Contaminants by Nanofiltration and Reverse Osmosis Membranes: Effects of Fouling, Modelling and Water Reuse

DISSERTATION
Submitted in fulfilment of the requirements of
the Board for Doctorates of Delft University of Technology
and of the Academic Board of the UNESCO-IHE Institute for Water Education
for the Degree of DOCTOR
to be defended in public
on Tuesday, 9 February 2010, at 12:30 hours
in Delft, the Netherlands

by

Victor Augusto YANGALI QUINTANILLA

BSc. and eng. in environmental engineering with distinction
National University of Engineering, Lima, Peru
Master of Science in municipal water and infrastructure
UNESCO-IHE, Delft, the Netherlands
born in Huancavelica, Peru

This dissertation has been approved by the supervisor
Prof. dr. G.L. Amy

Members of the Awarding Committee:

Chairman	Rector Magnificus TU Delft, the Netherlands
Prof. dr. A. Szöllösi-Nagy	Vice-Chairman, Rector UNESCO-IHE
Prof. dr. G.L. Amy	TU Delft / UNESCO-IHE, supervisor
Assoc. prof. dr. M. Kennedy	UNESCO-IHE, the Netherlands, cosupervisor
Prof. dr. J.C. van Dijk	TU Delft, the Netherlands
Prof. dr. B. van der Bruggen	Leuven Catholic University, Belgium
Prof. dr. K. Vairavamoorthy	University of Birmingham, United Kingdom
Dr. J.-C. Schrotter	Anjou Recherche – Veolia, France

CRC Press/Balkema is an imprint of the Taylor & Francis Group, an informa business

© 2010, Victor A. Yangali Quintanilla

Published by:
CRC Press/Balkema
PO Box 447, 2300 AK Leiden, the Netherlands
e-mail: Pub.NL@taylorandfrancis.com
www.crcpress.com - www.taylorandfrancis.co.uk - www.ba.balkema.nl

ISBN 978-0-415-58277-3 (Taylor & Francis Group)

Contents

Summary

The removal of organic micropollutants (1,4-dioxane, estrone, atrazine, bisphenol A, 17β-estradiol, 17α-ethynylestradiol, ibuprofen, naproxen, sulphamethoxazole, fenoprofen, ketoprofen, phenazone, carbamazepine, caffeine, acetaminophen, metronidazole and phenacetine) from synthetic water solutions and surface water was investigated in filtration experiments using NF membranes. A laboratory-scale flat sheet membrane unit was implemented to conduct the experiments. Two aromatic polyamide nanofiltration (NF) membranes (Dow-Filmtec) were utilized in laboratory-scale experiments. The membranes were NF-90 and NF-200, with molecular weight cut-offs of 200 and 300 Da, respectively. Additionally, other NF membranes (TS-80, Trisep; Desal HL, GE Osmonics; NE-90, Saehan; NF-70, Filmtec) and reverse osmosis (RO) membranes (XLE-440, LE-440, BW440, BW30LE, Dow-Filmtec; RE-BLR, Saehan; UTC-70, Toray; ES-20, Nitto Denko; AK, Osmonics; ESPA, ESNA, Hydranautics) were also studied taking into account data (of removal of micropollutants) generated during research conducted in other laboratory, pilot- and full-scale installations.

The results of the laboratory study showed organic solute rejections were not related to a single physicochemical descriptor. Organic solutes were classified into four groups: hydrophobic neutral, hydrophilic neutral, hydrophobic ionic and hydrophilic ionic. High rejections (90–99%) were observed for ionic compounds compared to neutral compounds (hydrophilic and hydrophobic, 20–99%). It was demonstrated that electrostatic repulsion between the negative charge of the ionic species of the solutes and the negative charge of the membrane surface was the main mechanism of rejection for ionic (charged) compounds. On the other hand, for hydrophilic neutral and hydrophobic neutral compounds, the rejections were mainly related to variables of size in the order of effective diameter $\approx$ length $>$ equivalent width $>$ molecular weight. Hydrophilic neutral compounds were less rejected by an NF membrane when their size (effective diameter, length or equivalent width) is less than or close to the pore size of the membrane ($<$ 1nm). By contrast, hydrophobic neutral compounds are influenced by log K_{ow} (hydrophobicity) contributing to partitioning when the pore size of the membrane is close to the size of the compound. Partitioning of a hydrophobic neutral compound (log $K_{ow} > 3$) will occur after adsorption of the solute into/onto the membrane until saturation; this effect was noticed only for BPA with NF-200 membranes. Adsorption of ionic and hydrophilic neutral compounds was less significant and did not interfere with the mechanisms of electrostatic repulsion and size exclusion. With respect to dipole moment, it appeared that high dipole moments decreased compound rejections for these compounds. However, it was observed that their small molecular size was more important than their high dipole moments.

Low fouling with sodium alginate (~15% flux decline) of the NF-200 membrane slightly decreased rejection of hydrophilic neutrals as well as hydrophilic and hydrophobic ionic compounds. Nevertheless, rejection of hydrophobic neutral compounds by NF-90 was not observed to be distinctly affected by alginate fouling, in fact, hydrophilic neutral compounds showed an increased rejection attributed to the domination of an enhanced sieving effect.

At a fouling degree equal to or less than 22% flux decline, mixed rejection results, either increased or decreased (±9%), were observed for neutral compounds by membranes fouled with sodium alginate, dextran and NOM. However, NOM fouling up to 50% flux decline increased rejection up to 14%, except for some compounds (caffeine, phenacetine). Ionic compounds showed increased rejection by membranes with 50% flux decline after NOM fouling (2–6%). Alginate fouling (22% flux decline) slightly increased rejection (0–4%). Dextran fouling (12% flux decline) and NOM fouling (15% flux decline) slightly increased rejection (1%). Direct filtration of feed water with NOM (after 35% flux decline) showed consistent increased rejection for all compounds (6–46%). Nonetheless, considering the particular use of one type of feed surface water, generalizing that other types of feed water containing different compositions of NOM will show similar results must be a cautious consideration.

Quantitative structure activity relationship (QSAR) analysis was used to quantify compound rejection by an NF membrane, in terms of organic compound physicochemical properties, and membrane characteristics. The QSAR model was constructed using internal experimental data. The model was internally validated using measures of goodness of fit and prediction, and subsequently, the model was externally validated with external data. The QSAR model identified that the most important variables that influence rejection of organic solutes were log K_{ow} (or log D), salt rejection, equivalent width and effective diameter. The model was in accordance with elucidated rejection mechanisms; rejection increased by size exclusion effects, and solute hydrophobicity decreased rejection due to adsorption and partitioning mechanisms. In addition, magnesium sulphate salt rejection incorporated steric hindrance and electrostatic repulsion effects that were related to the membrane structure and operating conditions. The use of molecular weight cut-off (MWCO) was acceptable for modelling purposes; however, NF membranes with a broad range of MWCO (pore size and distribution) may make it difficult to estimate the rejection of contaminants, thus magnesium sulphate salt rejection may be a more appropriate parameter for NF membranes.

The concept of QSAR for describing rejection was later improved and extended with the use of artificial neural network (ANN) models. Use of ANN models based on QSAR equations was an important tool to predict rejection of neutral organic compounds by NF and RO membranes with standard errors of estimation close to 5% and regression coefficients, R^2, of 0.97. It was finally demonstrated that size interactions between membrane and solutes were decisive for the removal of organic neutral solutes by NF and RO, and that hydrophobic interactions were less important. It was also demonstrated that magnesium sulphate salt rejection may be a possible lump parameter that defines size exclusion capability of neutral organic compounds by NF or RO membranes; however, it may only be valid in combination with solute descriptors and for a range of boundary experimental conditions.

Finally, a demonstration that NF, instead of RO, may be an effective barrier against pharmaceuticals, pesticides and endocrine disruptor compounds was discussed and developed. It raises the question of why RO is used in existing water reuse facilities when NF can be a more cost-effective and efficient technology able to tackle the problem of emerging organic contaminants. It was concluded that NF is a robust barrier for micropollutants when reduced O&M (operation and maintenance) costs are considered in a long-term project implementation. Savings of $0.02/m^3$ of treated water were estimated for NF compared to RO. It was also demonstrated that NF can comply with current regulations of water quality regarding pesticides removal, and its compliance with future regulations for pharmaceuticals and endocrine disrupters may be realistic. Specifically, when there is a potential presence of difficult to remove organic contaminants such as NDMA (N-Nitrosodimethylamine) and 1,4-dioxane, implementation in a WWTP (wastewater treatment plant) of additional processes to remove these compounds will help to reduce their presence during further water treatment with intentions of indirect reuse (e.g. after groundwater replenishment). In addition, other treatment techniques that are less expensive than advanced oxidation processes can help to remove 1,4-dioxane; and, biodegradation of NDMA can be achieved through aquifer recharge and recovery.

Samenvatting

In dit proefschrift werd de verwijdering (retentie) van verschillende organische microverontreinigingen (1,4-dioxaan, oestron, atrazine, bisphenol-A, 17β-estradiol, 17α-ethyniloestradiol, ibuprofen, naproxen, sulfamethoxazol, fenoprofen, ketoprofen, fenazon, carbamazepine, caffeine, acetaminophen, metronidazole, phenacetine) uit verschillende modelwaters en oppervlaktewater met nanofiltratie (NF) en omgekeerde osmose (RO) membranen onderzocht met behulp van filtratie-experimenten. Hiervoor werd gebruik gemaakt van een labschaal-installatie met vlakke membraansheets. Twee verschillende NF membranen (van Dow-Filmtec, met een toplaag van aromatisch polyamide) werden gebruikt in de experimenten op lab-schaal: NF-90 en NF-200, met respectievelijke *molecular weight cut-off*-waardes van 200 en 300 Da. Daarnaast werden ook vroeger bekomen retenties van organische microverontreinigingen met andere NF membranen (TS-80, Trisep; Desal HL, GE Osmonics; NE-90, Saehan; NF-70, Filmtec) en RO membranen (XLE-440, LE-440, BW440, BW30LE, Dow-Filmtec; RE-BLR, Saehan; UTC-70, Toray; ES-20, Nitto Denko; AK, Osmonics; ESPA, ESNA, Hydranautics) gebruikt voor het opstellen van modellen in deze studie. Resultaten van zowel lab-, piloot- als full-scale installaties werden gebruikt.

De resultaten van de proeven op labschaal toonden aan dat verwijdering van organische verontreinigingen niet volledig te relateren valt aan één enkele fysisch-chemische eigenschap van de stoffen. De organische verontreinigingen werden opgedeeld in vier verschillende groepen: hydrofobe ongeladen, hydrofiele ongeladen, hydrofobe geladen en hydrofiele geladen verbindingen. De hoogste verwijderingspercentages (90-99%) werden waargenomen voor de geladen stoffen; voor neutrale stoffen (zowel hydrofiele als hydrofobe) was de verwijdering lager (20-99%). Er kon aangetoond worden dat de hogere verwijdering van negatief geladen stoffen te wijten is aan electrostatische afstoting tussen de negatieve lading van de organische stof en de negatieve membraanlading. Voor de hydrofiele en hydrofobe neutrale stoffen, daarentegen, was de verwijdering (in aflopende volgorde van belangrijkheid) afhankelijk van de effectieve diameter; de moleculaire lengte (length); de equivalente breedte (equivalent width) en het molecuulgewicht (molecular weight) van de stoffen. Hydrofiele neutrale stoffen werden minder goed tegengehouden door NF membranen wanneer hun grootte (als effectieve diameter, lengte of equivalente breedte) kleiner was dan de grootte van de membraanporiën (< 1nm). Dit in tegenstelling tot hydrofobe neutrale stoffen, waarvan de verwijdering bovendien afhankelijk was van de log K_{ow}-waarde (die de hydrofobiciteit van de stof aangeeft). Een hoge log K_{ow}-waarde draagt bij aan het makkelijker "binnendringen" *(partitioneren)* van een stof in het membraan als zijn grootte ongeveer gelijk is aan de grootte van de membraanporiën. Hydrofobe (log K_{ow} > 3) ongeladen organische stoffen kunnen adsorberen aan het membraanoppervlak en in de membraanporiën. Eens het membraan verzadigd is, zorgt de hoge hydrofobiciteit van de stof bovendien nog steeds voor een verhoogde *partitionering* (wat erin resulteert dat hydrofobe stoffen makkelijker in het membraan kunnen binnendringen). Dit effec werd bijvoorbeeld waargenomen voor de stof Bisphenol-A in combinatie met het NF-200 membraan.

Adsorptie van geladen stoffen en van ongeladen hydrofiele stoffen was minder significant dan bij hydrofobe ongeladen stoffen en had daar ook geen invloed op de verwijderingsmechanismes (voornamelijk electrostatische afstoting en scheiding op basis van grootte (~het zeefeffect)). Het dipoolmoment van de organische stoffen had ook (beperkte) invloed op de verwijdering: voor stoffen met hoge dipoolmomenten bleek de retentie lager te zijn. Er moet echter opgemerkt worden dat de stoffen met een hoog dipoolmoment vaak ook de kleinste moleculen waren, waardoor de kleine moleculaire afmetingen van deze stoffen soms meer effect hadden op de verwijdering dan het hoge dipoolmoment.

Beperkte vervuiling (~15% fluxdaling) van het NF-200 membraan door alginaat veroorzaakte een lichte daling van de retentie van ongeladen hydrofiele en geladen hydrofiele en hydrofobe stoffen. Voor het NF-90 membraan, daarentegen, had vervuiling met alginaat geen invloed op de retentie van ongeladen hydrofobe stoffen. Voor ongeladen hydrofiele stoffen nam de retentie toe door het toegenomen zeefeffect van de vervuilingslaag.

Voor membranen vervuild met dextraan en natuurlijk organisch materiaal (NOM) werden, bij een fluxdaling kleiner of gelijk aan 22%, verschillende invloeden van de vervuiling op de retentie van ongeladen organische stoffen waargenomen: zowel stijgingen als dalingen van de retenties (±9%) werden waargenomen. Bij zware NOM vervuiling (fluxdaling van 50%), daarentegen, stegen de retenties van de ongeladen organische stoffen (tot 14% hoger), behalve voor enkele uitzonderingen (caffeine en phenacetine). Voor geladen stoffen werd een ander beeld waargenomen: bij vervuiling door dextraan (12% fluxdaling) en beperkte NOM vervuiling (15% fluxdaling) stegen de retenties ongeveer 1%. Bij zware NOM vervuiling (50% fluxdaling) stegen de retenties iets meer (2-6%).

Directe nanofiltratie van een natuurlijk watertype dat NOM bevatte (en resulteerde in 35% fluxdaling) zorgde voor een toegenomen verwijdering van alle organische microverontreinigingen (6 tot 46% hoger). Vertaling van de resultaten met dit ene type oppervlaktewater naar andere watertypes (die andere types NOM bevatten) is echter moeilijk en er moet omzichtig mee omgesprongen worden.

Ook in dit proefschrift werden structuur-activiteitsrelaties (QSAR) gebruikt om retenties van organische stoffen met NF membranen te modelleren als functie van de fysisch-chemisch eigenschappen van de organische stoffen en het membraan. De QSAR modellen in dit proefschrift werden opgesteld met experimentele data die intern beschikbaar was. Het model werd eerst intern getoetst, met behulp van statische parameters als de regressiecoefficient, de goedheid van de fit, en op basis van de nauwkeurigheid van voorspelling. Nadien werd het model ook getoetst met een externe database. Uit het QSAR model kwam naar voor dat de meest belangrijke parameters voor retentie de log K_{ow} (of log D) en effectieve diameter en equivalente breedte van de stof waren, in combinatie met de zoutretentie van het membraan. De conclusies uit het model stemden dus overeen met de conclusies uit de experimenten: retentie neemt toe door een toename in het zeefeffect (m.a.w. grotere stoffen worden beter tegengehouden dan kleinere stoffen) en retentie neemt af door een toename in adsorptie aan en *partitionering* in het membraan (bij hogere log K_{ow}). Dat retentie van organische stoffen afhankelijk is van zoutretentie van het membraan is ook logisch: retentie van magnesiumsulfaat is zowel afhankelijk van scheiding op basis van grootte als van electrostatische afstotingseffecten (beiden zijn afhankelijk van de membraanstructuur). Daarnaast is zoutretentie, net als retentie van organische stoffen, afhankelijk van de procesbedrijfsvoering. In sommige gevallen was het ook mogelijk geweest de Molecular Weight Cut-Off (MWCO) waarde van de membranen te gebruiken in de modellen. Omdat sommige NF membranen echter geen nauwkeurig gedefinieerde MWCO hebben (omdat er een poriegrootte-verdeling is i.p.v. een nauwkeurig gedefinieerde poriegrootte), lijkt zoutretentie een betere parameter te zijn dan MWCO voor modelleringsdoeleinden.

De opgestelde QSAR modellen voor retentie zijn in een volgend stadium ook verbeterd en uitgebreid met neurale netwerk-modellen (artificial neural networks (ANN)). De ANN modellen, gebaseerd op de QSAR-vergelijkingen, bleken in staat om retenties van ongeladen organische stoffen op NF en RO membranen nauwkeurig te voorspellen (standaardafwijkingen rond de 5% en regressie-coefficiënten (R^2) rond de 0,97). De ANN modellen bevestigden bovendien dat scheiding op basis van grootte het belangrijkste mechanisme was voor retentie, en dat hydrofobe interacties minder belangrijk waren. Er werd ook nogmaals aangetoond dat zoutretentie (bv. van magnesiumsulfaat) als parameter kan gebruikt worden om scheiding op basis van grootte te beschrijven. Er moet echter opgemerkt worden dat zoutretentie in sommige gevallen enkel bruikbaar is in combinatie met bepaalde stofeigenschappen en voor een bepaalde set van experimentele condities.

Tenslotte werd er ook aandacht besteed aan het gebruik van NF, in plaats van RO, als effectieve barrière tegen geneesmiddelen, bestrijdingsmiddelen en hormoonverstorende stoffen. Men kan zich de vraag stellen waarom meestal RO membranen gebruikt worden in bestaande installaties voor afvalwaterhergebruik, terwijl NF membranen even efficiënt én bovendien meer economisch kunnen zijn om het probleem van de organische microverontreinigingen aan te pakken. In dit proefschrift werd geconcludeerd dat NF een economisch haalbare én robuuste barrière kan vormen voor microverontreinigingen als de verminderde operationele en managementskosten (t.ov. RO) in de projectieberekeningen meegenomen worden. In dit proefschrift werd geschat dat het gebruik van NF in plaats van RO een kostenbesparing van 0,02$ per m³ behandeld water kan opleveren. Er werd ook aangetoond dat met NF membranen aan de huidige wetgeving voor verwijdering van bestrijdingsmiddelen voldaan kan worden, en dat het halen van eventuele toekomstige normen voor geneesmiddelen en hormoonverstorende stoffen realistisch is. Indien stoffen in het afvalwater voorkomen waarvan geweten is dat ze moeilijk te verwijderen zijn door NF/RO (bv. NDMA en 1,4-dioxaan), moeten deze door andere processen in de afvalwaterzuivering en/of drinkwaterzuivering verwijderd worden. Zo kunnen andere technieken, die goedkoper zijn dan geavanceerde oxidatietechnieken gebruikt worden om 1,4-dioxaan te verwijderen, en kan NDMA verwijderd worden door gebruik te maken van biologische afbraak tijdens bodempassage.

Chapter 1

Introduction

1.1   Background

Now and in the future, the ever-growing demand for drinking water will lead many cities to implement indirect water reuse programs, where wastewater effluent becomes part of the drinking water sources. Pollution of those sources with emerging contaminants such as endocrine disrupting compounds (EDCs), pharmaceutically active compounds (PhACs) and personal care products (PCPs) is a fact known worldwide. The presence of these emerging contaminants is of increasing concern in drinking water treatment plants that recycle wastewater effluents or use wastewater-contaminated surface waters. Although the risks of PhACs, PCPs and EDCs polluting sources of water are partly recognized, interpretation of consequences are controversial; thus, the future effects of altered water with trace contaminants remains uncertain and may constitute a point of concern for human beings when potable water consumption is involved. Therefore, many drinking water utilities target as an important goal high-quality drinking water production to lessen quality considerations that may arise from the consumers.

Nonetheless, in search of precautionary measures against the unforeseen consequences that those compounds may cause, the study of their removal through membrane treatment is presently of great scientific interest for the future, as water resources become scarce and demands for recycled water increase. Mainly reverse osmosis (RO) has been demonstrated to be an appropriate technology for removing a large number of emerging organic contaminants, but nanofiltration (NF) also constitutes a good option although believed to have certain limitations. Nonetheless, the performance of RO and NF can be questioned because there are limited tools that optimise quantification of the removal of contaminants. This is mainly because the achievement of a fundamental (theoretical) understanding to describe interactions occurring between membranes and organic solutes to quantify their performance can be a difficult task. Therefore, the use of alternative techniques (e.g. multivariate data analysis) can be more practical and effective to understand and model the separation of organic contaminants by membranes.

1.2   Objectives and research questions

The main objectives of this thesis are

   o   Identify and understand interactions occurring between organic solutes and membranes that influence rejection during membrane filtration.

- ○ Demonstrate that modified properties of fouled membranes when compared to clean ones may result in improved, diminished or equal removal of organic solutes.

- ○ Define models that may describe or predict removal of organic compounds based on solute properties and membrane characteristics.

- ○ Demonstrate that NF may be efficient for removal of pharmaceuticals, pesticides and endocrine disrupters.

The following research questions were formulated based on an analysis of existing knowledge revealed by former investigations.

- ○ What are the main solute properties that contribute to a better understanding of solute rejection by membranes during filtration?

- ○ What are the membrane characteristics that influence the rejection of different groups of solutes?

- ○ How does the rejection occur? What types of interactions/mechanisms occur during filtration?

- ○ Does fouling alter membrane properties and, hence, solute rejection?

- ○ How can rejection trends be correlated to mathematical models? Which models can better describe experimental results?

1.3   Organization of thesis

This thesis contains eight chapters. The main chapters (Chapters 3–7) are based on peer-reviewed scientific journal publications and conference papers and proceedings. Chapter 2 presents an introductory theoretical background of important concepts, definitions and methods used throughout the thesis. Chapter 3 deals with experimental work carried out using NF membranes (clean and fouled with sodium alginate) for filtration of pharmaceuticals and endocrine disrupters. Chapter 3 also explains the effects of the surrogate foulant sodium alginate on rejection of organic compounds and changes in membrane characteristics due to adsorption of foulants. This chapter also describes concentration polarization of organic solutes in NF membranes and compares rejection changes produced by different hydrodynamic conditions.

Chapter 4 deals with NF experiments with organic compounds by considering the fouling effects of natural organic matter (NOM) and surrogate foulants (dextran, sodium alginate). Chapter 4 has been separated from Chapter 3 because they differ in experimental approach. To explicate, additional organic compounds, more membrane foulants, different feed water synthetic solutions, different hydrodynamic experimental conditions and different degrees of fouling and flux decline were used in Chapter 4. Also in this chapter, surrogate and NOM foulants of membranes are characterised with ATR-FTIR spectra, contact angles, surface charge measurements and salt rejection tests. Moreover, physicochemical properties (molecular weight, molecular length, effective diameter, equivalent width, octanol-water partition coefficients, dipole moments) of organic compounds are used to compare similar or different removals achieved by clean and fouled membranes.

Chapter 5 develops a quantitative structure-activity relationship (QSAR) model for predicting the rejection of pharmaceuticals, endocrine disrupters, pesticides and other organic compounds by NF membranes. Principal component analysis, partial least-square regression and multiple linear regressions were used to find a general QSAR prediction model that combines interactions between membrane characteristics, filtration operating conditions and compound properties. An internal database was used to produce and validate the QSAR model. Subsequently, the model was validated with an external database of compound rejections.

Considering the good modelling results obtained in Chapter 5, Chapter 6 deals with the development of an extended model applicable to NF and RO. Prediction of the rejection of neutral organic compounds by polyamide NF and RO membranes was evaluated using artificial neural network (ANN) models. The ANN models were developed based on QSAR equations that defined an appropriate set of solute and membrane variables able to represent and describe rejection. An internal database in combination with data collected from the literature was used to produce the ANN models.

Chapter 7 uses results obtained in this study and other existing information to demonstrate that NF is an effective barrier against pharmaceuticals and endocrine disrupters. It raises the question of why RO is used in existing water reuse facilities when NF may be a more cost-effective and efficient technology able to tackle the problem of emerging organic contaminants. Sufficient support information and, more importantly, experimental and current practice results provide proof that NF should be considered instead of RO in future water reuse projects.

Chapter 2

Theoretical background

2.1   Regulation of emerging contaminants

Nowadays, the presence of PhACs, PCPs and EDCs in drinking water samples remains unregulated. Article 16 of the Water Framework Directive 2000/60/EC (European Parliament, 2000) sets out a strategy for dealing with the chemical pollution of water; the first step of this strategy is a list of priority substances adopted in the Decision 2455/2001/EC, identifying 33 substances of priority concern at the community level (European Parliament, 2001). A more recent proposal aims to ensure a high level of protection against risks to or via the aquatic environment stemming from these 33 priority substances and certain other pollutants by setting environmental quality standards (European Parliament, 2006). In consequence, programs of excellent water quality in the Netherlands have been implemented.

The vision of the Dutch water companies is that there is a need for constant improved water quality to keep the Dutch inhabitants' confidence up for their direct consumption of high-quality tap water instead of bottled water. Based on that vision, a large number of research projects have been started in the European Union (EU). EU projects aimed at high-quality potable water are the Delft Cluster and Techneau projects; both include work packages that investigate options to deal with the presence of emerging organic contaminants in water. The Q21-PODW (Production of Outstanding Drinking Water for the 21st century) was part of one EU project started at Delft University of Technology in 2006.

In order to address the potential environmental and health impacts of endocrine disruption, the European Commission adopted a Communication to the Council and European Parliament, entitled "Community Strategy for Endocrine Disrupters" in December 1999 (European Parliament, 1999). This strategy sets out a number of actions relating to the identification of substances, monitoring, research, international coordination and communication to the public. On 26 October 2000, the European Parliament adopted a resolution on endocrine disrupters, emphasising the application of the precautionary principle. The epidemiological evidence of potential relationships between exposure to chemical substances and endocrine disruption is a general cause for concern. Although a considerable amount of research is still required to ascertain the scope and seriousness of endocrine disruption, including confirmation of epidemiological results, it is essential that the Commission adopt a strategy that takes into account the current concern on the basis of the precautionary principle (Ibid).

REACH is the new European regulation (EC) No 1907/2006 on chemical substances. REACH stands for Registration, Evaluation and Authorisation of CHemicals. REACH allows the evaluation of substances of concern and foresees an authorisation system for the use of substances of very high concern. This applies to substances that cause cancer, infertility, genetic mutations or birth defects, and to those which are persistent and accumulate in the environment. REACH requires a registration, over a period of 11 years, of some 30,000 chemical substances. The registration process requires the manufacturers and importers to generate data for all chemicals substances produced or imported into the EU above one tonne per year. The registrants must also identify appropriate risk management measures and communicate them to the users. The supervision of compliance with REACH comes under the responsibility of each member state. Belgium and the Netherlands have already enforced its fulfillment. In the Netherlands, the Dutch Labour Inspectorate, the Ministry of Housing, Spatial Planning and the Environment Inspectorate (VROM Inspectorate) and the Food and Consumer Product Safety Authority are dealing with the enforcement of REACH at different levels. The Labour Inspectorate supervises the professional users of substances and preparations. The Food and Consumer Product Safety Authority supervises producers, importers and dealers in preparations and items for consumers. The VROM Inspectorate supervises producers, importers and dealers of substances, preparations and items for professional use.

In the United States, the Safe Drinking Water Act (1974) required the Environmental Protection Agency (EPA) to establish maximum levels for various drinking water contaminants including some pesticides known to have endocrine disruptive activity. According to Snyder et al. (2003), endocrine disruption was not specifically named in any United States legislation until 1995. Amendments to the Safe Drinking Water Act mandated that chemicals and formulations be screened for potential endocrine activity before they are manufactured or used in certain processes where drinking water and/or food could become contaminated. Thus, the EPA was required to develop a screening program, using appropriate, validated test systems and other scientifically relevant information to determine whether certain substances may have an endocrine effect on wildlife and humans (EPA, 1998). Nonetheless, the legislation regulated only those industries producing or using raw chemicals, and not the water industry. As a result, legislation will have no immediate effect on water and wastewater treatment regulations. There are currently no federal regulations for pharmaceuticals in drinking or natural waters. The Food and Drug Administration (FDA) requires ecological testing and evaluation of a pharmaceutical only if an environmental concentration in water or soil is expected to exceed 1mg/L or 100mg/kg, respectively (FDA, 1998).

2.2   Presence of EDCs, PhACs and PCPs in water sources

2.2.1   Source and threat

According to the International Program for Chemical Safety, an endocrine disrupter is an exogenous substance or mixture that alters function(s) of the endocrine system and consequently causes adverse health effects in an intact organism, or its progeny or (sub) populations (European Parliament, 1999). More than 70,000 chemicals are believed to be EDCs (endocrine disrupting compounds). The main evidence suggesting that exposure to environmental chemicals can lead to disruption of endocrine function comes from changes seen in a number of wildlife species. Many articles have reported threats to health and reproductive biology in animal populations; some examples include an increased uterus growth in rats (Bicknell et al., 1995), changes in the gonads of alligators (Guillette et al., 1994). Other effects suggested as being related to endocrine disruption have been reported in molluscs, crustacea, fish, reptiles, birds and mammals in various parts of the world (Sumpter and Johnson, 2005). However, there is limited evidence of adverse endocrine-mediated effects in humans that have followed either intentional or accidental exposure to high levels of particular chemicals. The clearest example of an endocrine disrupter in humans is diethylstilbestrol, a synthetic oestrogen prescribed in the 1950s and 1960s to five million pregnant women for the prevention of spontaneous abortion; it was found that some of the children who had been exposed to this drug in the uterus had developmental abnormalities (Sawyer et al., 2003).

Pharmaceuticals are defined as chemicals used for the treatment or prevention of illness. As such, they can range from compounds used for cancer treatment and birth control, to antibiotics used to combat infection, to compounds used to relieve pain (e.g. aspirin and ibuprofen). Pharmaceuticals are also used in veterinary health care, e.g. antibiotics and growth hormones (Sawyer et al., 2003). From an environmental point of view, pharmaceuticals are a class of emerging environmental contaminants that are extensively and increasingly being used in human and veterinary medicine. Pharmaceuticals are designed to have a specific mode of action and many of them develop some persistence in the body. These features, among others, make it necessary for pharmaceuticals to be evaluated for potential effects on aquatic flora and fauna (Fent et al., 2006).

Very little is known with respect to the effect of PhACs on human and wildlife health. However, there is potential evidence of adverse effects because most of these compounds are lipophilic and their activity is slow to decay, meaning that they remain pharmaceutically active for an extended period of time. Thus, bio-concentration is possible. While the concentration of individual PhACs in water supplies is low (generally less than 0.5μg/L with many in the 100ng/L range), the presence of numerous drugs with similar modes of action could lead to measurable effects, especially in pregnant women and babies. Finally, exposure can be chronic because PhACs are continually introduced into the environment via human wastewater treatment, direct discharges, livestock production, aquaculture, hospitals and similar environments.

Personal care products (PCPs) comprise products such as soaps, shampoos, conditioners, toothpastes, fragrances, disinfectants and antiseptics, sunscreens and varied cosmetics. After use of PCPs, the ingredients that are in them enter wastewater streams. Normally, most wastewater treatment plants are not sufficiently implemented to be capable of removing organic solutes resulting from PCPs and, therefore, they will bio-accumulate in the environment (Ternes et al., 2004; Ellis, 2006). Moreover, the effects of organic solutes derived from PCPs' ingredients going into aquatic environments are not well understood (Boyd et al., 2003; Wilson et al., 2003). In summary, pharmaceuticals, personal care products, endocrine disrupters, pesticides and other organic contaminants follow the cycle presented in Fig. 2.1. A municipal waste water treatment plant (WWTP) does not always exist in developing countries where direct wastewater discharge may occur. Moreover, depending on the processes and technologies implemented in the WWTP, the removal efficiencies of contaminants widely vary (Ternes et al., 2004); the same applies for drinking water treatment plants (WTP).

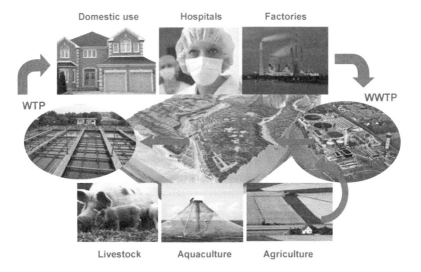

Fig. 2.1: Routes and cycle of emerging organic contaminants in the water cycle

2.2.2   Occurrence

The fate of EDCs, PhACs and PCPs in the environment has raised the interest of scientists because the accumulation of non-degradable xenobiotics may result in environmentally significant concentrations with unknown effects. Recent studies have demonstrated the presence of residues of commonly used pharmaceuticals in the treated waters from sewage treatment plants and in aquatic environments in Europe; the concentration level for pharmaceuticals ranged from nanograms to micrograms per litre (Ternes, 1998; Hirsch et al., 1999; Ollers et al., 2001). The concentrations of PhACs, PCPs and EDCs measured in the environment are in the range of nanograms per litre (ng/L) to a few micrograms per litre (μg/L) (Clara et al., 2005).

A review of research data of investigations carried out in Austria, Brazil, Canada, Croatia, England, Germany, Greece, Italy, Spain, Switzerland, the Netherlands, and the United States of America (USA) revealed that more than 80 compounds, including pharmaceuticals and several drug metabolites, were detected in those aquatic environments (Heberer, 2002). Several PhACs from various prescription classes have been found at concentrations up to the μg/L-level in sewage influent and effluent samples and also in several surface waters located downstream from WWTPs. The studies show that some PhACs originating from human therapy are not completely eliminated in the WWTP and, therefore, are discharged as contaminants into receiving waters. Under groundwater recharge conditions, polar PhACs such as clofibric acid, carbamazepine, primidone or iodinated contrast agents can leach through the subsoil and have also been detected in several groundwater samples in Germany (Heberer, 2002).

After the first nationwide reconnaissance across 30 states in the USA during 1999 and 2000, 95 organic wastewater contaminants (pharmaceuticals, hormones and others) were identified in water samples from a network of 139 streams (Kolpin et al., 2002). Boyd et al. found concentrations of nine pharmaceuticals and personal care products (PCPs) in samples from two surface water bodies, a sewage treatment plant effluent and in various stages of a drinking water treatment plant in Louisiana, USA, as well as from one surface water body, a drinking water treatment plant and a pilot plant in Ontario, Canada (Boyd et al., 2003). The drug residues of lipid regulators, anti-inflammatories and some drug metabolites were identified in raw sewage, treated wastewater and river water in the state of Rio de Janeiro, Brazil; the median concentrations in the effluents of sewage treatment plants of drugs investigated ranged from 0.1 to 1 μg/L (Stumpf et al., 1999).

It is a fact that the occurrence of emerging contaminants is an international topic; its presence can be expected in rivers, lakes, wells and even drinking water. Maximum concentrations of pharmaceuticals reported in WWTP effluents, surface water, groundwater and drinking water are presented in Table 2.1. For the particular case of the Netherlands and Flanders (Belgium), Table 2.2 displays the concentrations of emerging contaminants in surface and drinking water.

Table 2.1: Maximum concentrations of pharmaceuticals in water

Compound	WWTP effluent (µg/L)	Surface water (µg/L)	Ground-water (µg/L)	Drinking water (µg/L)
Acetaminophen	6.0	10		
Diclofenac	2.5	1.2		0.006
Ibuprofen	85	2.7		0.003
Ketoprofen	0.38	0.12		
Naproxen	3.5	0.4		
Oxytetracycline		0.34		
Tetracycline		1.0		
Ciprofloxacin	0.13	0.07	0.02	
Carbamazepine	6.3	1.1	1.1	0.258
Metoprolol	2.2	2.2		
Clofibric acid	1.6	0.55	4.0	0.270
Iohexol	7.0	0.5		
Iopromide	20	4.0		0.086
17α-ethynilestradiol	0.003	0.83		
Ifosfamide	2.9			
Salbutamol		0.04		

Adapted from Weinberg et al., 2008

Table 2.2: Maximum concentrations of organic contaminants in water in the Netherlands, Flanders (Belgium) and the European Union

Compounds	Surface water (µg/L)		Drinking water (µg/L)	
	Flanders	Netherlands	EU	Netherlands
Endocrine disrupters				
17beta-estradiol	0.002	0.001	0.002	<0.0004
17α-ethynilestradiol	n.a.	0.004	<1	<0.0004
Estrone	0.022	0.003	0.022	<0.0004
Industrial chemicals				
Bisphenol A	0.580	22.0	22.0	<0.01
Phthalates	10.3	200	200	2.1
PCB	<0.007	0.020	0.08	<0.01
Nonylphenolpolyethoxylates	n.a.	2.6	2.6	1.5
MTBE	n.a.	62.0	62	<1
NDMA	n.a.	<10	<10	0.002
Pesticides				
Atrazine	13.0	0.4	13	0.03
Simazine	19.0	0.05	19	<0.01
DDT	<d.l.	<10	1.4	<0.01
Pharmaceuticals				
Sulphamethoxazole	n.a.	0.09	1.7	0.04
Carbamazepine	n.a.	0.5	2.0	0.09
Ibuprofen	n.a.	0.12	1.0	0.02
Iopamidol	n.a.	0.47	.47	0.07
Amidotrizoic acid	n.a.	0.29	.30	0.08

Adapted from Verliefde et al., 2007; n.a. data not available, d.l. detection limit.

In summary, diverse organic compounds such as EDCs, PhACs, pesticides, solvents and PCPs are present as contaminants at low concentrations in surface water, sewage treatment plant effluents, stages of drinking water treatment plants, and even at trace levels in finished drinking water (Hirsch et al., 1999; Kolpin et al., 2002; Boyd et al., 2003; Verliefde et al., 2007). The possible effects on human health and aquatic organisms, associated with the presence or consumption of water containing low concentrations of micropollutants, are documented in toxicology studies but are not fully known (Webb et al., 2003; Cleuvers, 2004; Jones et al., 2005; Fent et al., 2006).

The impact of domestic and industrial wastewater discharges, either treated or untreated, in surface water allocated to the production of drinking water is of increasing concern due to the introduction of various emerging organic contaminants in the water cycle that may result in alarming consequences, e.g. reduction of egg production in fish, feminization of fish, and possible growth inhibition of human embryonic cells (Ternes et al., 2004; Pomati et al., 2006; Jackson and Sutton, 2008; Thorpe et al., 2009). PhACs, EDCs, organic compounds derived from PCPs and other organic compounds discharged by diverse industries are either only moderately removed or not removed at all during wastewater treatment and afterwards during conventional drinking water treatment (Snyder et al., 2003; Vieno et al., 2006; Zorita et al., 2009). As a result, organic compounds have been detected in many surface waters in the Netherlands (Verliefde et al., 2007), Italy (Castiglioni et al., 2006), Greece (Stasinakis et al., 2008) and in China (Peng et al., 2008), to name a few. Moreover, a recent study reported the presence of pharmaceuticals, EDCs and other unregulated organic contaminants in USA drinking water (Benotti et al., 2008).

2.3   Physicochemical properties of organic solutes

2.3.1   Molecular weight, molar volume and size

The molecular weight (MW) is the molecular mass of a compound. MW can be used for the rejection prediction of non-charged and non-polar compounds in ultra low pressure membrane applications (Ozaki and Li, 2002). Often molecular weight is the most used parameter reflecting molecular size, however, it is not a direct measure of size. Other possible parameters are the effective diameter (Van der Bruggen et al., 2000), which projects the molecule onto a membrane surface, and the molecular width (Kiso et al., 1992).

Another two studies by Kiso et al. compared the use of geometrical descriptors of molecular length and molecular widths with the Stokes radius of the same molecules and reported a high correlation between the two (Kiso et al., 2000 and 2001a). Other size descriptors are molar volume (MV), molecular length, molecular width, molecular depth and equivalent molecular width. The molecular length (Fig. 2.2) is defined as the distance between the two most distant atoms. The molecular width and molecular depth (width > depth) are measured by projecting the molecule onto the plane perpendicular to the length axis; the equivalent molecular width is defined as the geometric mean of width and depth (Santos et al., 2006). Geometrical size descriptors are calculated after optimization geometry of a molecule from the interaction of conformational analysis and energy minimization with a semi-empirical quantum chemistry algorithm.

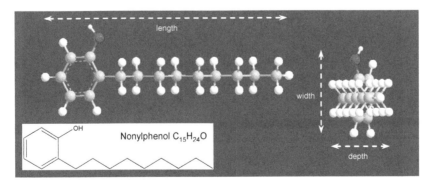

Fig. 2.2: Geometrical size descriptors of a compound

The effective diameter (Van der Bruggen et al., 2000) is defined as a mean height in projection. An axis is defined by the line between the two most distant atoms; the solute is projected onto the plane perpendicular to the axis and a cylinder is formed (Figure 2.3). The axis of the cylinder forms an angle α with the surface of the membrane. The cylinder is projected on the membrane surface and is calculated as

$$Height\ in\ projection = a\cos\alpha + b\sin\alpha \qquad (2.1)$$

where a = height of the cylinder, and b = diameter of the cylynder.

After assuming that the probability of an arbitrary angle α is proportional to the surface of a spherical shell, which results in a propability distribution $p(\alpha) = cos\ \alpha$. The mean height in projection (the effective diameter) can be calculated as

$$Effective\ diameter = \frac{1}{\pi/2}\int_0^{\pi/2}(a\cos\alpha + b\sin\alpha)p(\alpha)d\alpha$$

$$Effective\ diameter = \frac{a}{2} + \frac{b}{\pi} \qquad (2.2)$$

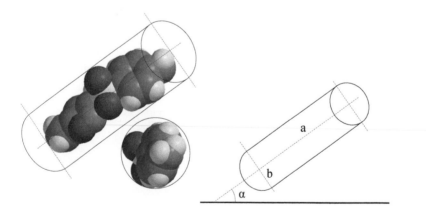

Fig. 2.3: Drawing of solute in a cylinder to define an effective diameter

2.3.2   Solubility

The solubility of a given compound in water reflects its affinity for water. Thus, the more soluble it is in water, the more efficient the compound remains in the aqueous solution, and less adsorption on the membrane surface takes place. Therefore, water solubility might be the first indication of the effect of a compound on its passage through a membrane.

2.3.3   Acid dissociation constant

The acid dissociation constant (K_a) is an equilibrium constant that measures the ability of a Brönsted acid to donate a proton to a specific reference base. The greater the value of K_a is, the stronger the acid. In dilute aqueous solutions, water is the reference base. The pK_a is defined as $-\log K_a$. The base dissociation constant (K_b) is an equilibrium constant that characterises the ability of a Brönsted base to accept a proton from a reference acid. The pK_b is defined as $-\log K_b$. The product of the aqueous acid and base dissociation constants of conjugate pairs is equal to K_w, the ionization constant for water.

Acid dissociation constants are equilibrium constants that define 50% of the formation of dissociated species. According to the pH of a solution, as shown in Fig. 2.4, the percentage of dissociated species of a solute can be determined and classified as ionic (negatively or positively charged fractions) or non-ionic (neutral fraction).

2.3.4   Dipole moment

The dipole moment is defined as a vectorial property of individual bonds or entire molecules that characterises their polarity. A molecular dipole moment is the dipole moment of the molecule taken as a whole. It is a good indicator of a molecule's overall polarity. The value of the molecular dipole moment is equal to the vector sum of the individual bond dipole moments (see Fig. 2.5). This vector sum reflects both the magnitude and the direction of each individual bond dipole moment. Therefore, such a molecule acts as a dipole and tends to become aligned in an electrical field. The bond dipole moment μ is obtained by multiplying the charge at either atom (pole) q (in electrostatic units or esu) by the distance d (in centimetres) between the atoms (poles): $q \cdot d\_\mu$ (in esu-cm). Dipole moments are usually expressed in Debye units, equal to 10^{-18} esu-cm (Orchin and University of Cincinnati, Dept. of Chemistry, 2005).

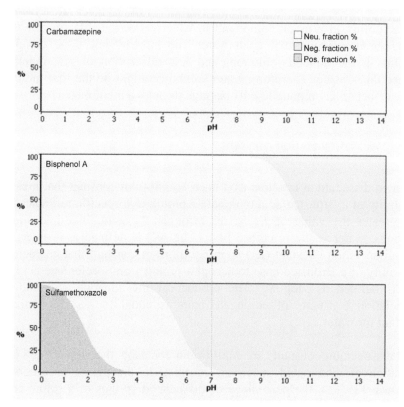

Fig. 2.4: Species of solutes present in water over a range of pH

Adapted from ADME/Tox Web Software

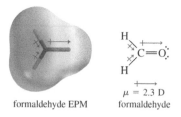

formaldehyde EPM   formaldehyde

Fig. 2.5: Dipole moment of formaldehyde with electrostatic potential map (EPM)

(Wade, 2003)

2.3.5   Octanol-water partition coefficients

The octanol-water partition coefficient (K_{ow}) is often used to describe hydrophobicity. Among partition effects of unionised (neutral or non-ionic) organic compounds in various solvent-water mixtures, the partition coefficients in octanol-water mixtures have received attention because of the observed correlations between the octanol-water partition coefficients and partition effects with natural organic substances (Chiou, 2002).

The octanol-water partition coefficient is a measure of the equilibrium concentration of a compound between octanol and water. The octanol-water partition coefficient, used in logarithmic form as log K_{ow} = log (C_o/C_w), with C_o = concentration of the compound in the octanol phase, and C_w = concentration of the unionised compound in the water phase, is a measure of the hydrophobicity of a given compound.

The octanol-water distribution coefficient, D, is the ratio of the equilibrium concentrations of all species (unionised and ionised) of a molecule in octanol to the same species in the water phase at a given temperature. It differs from K_{ow} in that ionised species are considered, as well as the neutral form of the molecule. D is defined as $(C_i)_{oct}$ / $(C_i)_{aq}$ where i correspond to ionised and un-ionised species and the subscripts 'oct' and 'aq' stand for the octanol and the water phase, respectively. D is expressed as log D in logarithmic form. If a compound contains one or more ionisable groups, it may exist in solution as a mixture of different ionic forms. The composition of this mixture depends strongly on pH. Log K_{ow} is defined only for neutral species; the partition coefficient for partially ionised mixtures or the effective partition coefficient for dissociative systems gives the correct description of the complex partitioning equilibrium (Sangster, 1997). Since log D reflects the true behaviour of an ionisable compound in a solution at a given pH value or range, this property is useful in evaluating pH-dependent properties and adsorption processes.

Figure 2.6 shows a clarification of the concepts of log D and log K_{ow}. Compounds with log $K_{ow} \geq 2$ are referred to as hydrophobic (HB); and those with log $K_{ow} < 2$ are hydrophilic (HL). This classification was based on an early reference (Connell, 1990). However, a log K_{ow} (log D) higher or equal than 3 can also be used to refer an organic compound as hydrophobic. The effect of unionised and ionised species in log D for bisphenol A and sulfamethoxazole is clearly observed after comparing Fig. 2.4 and 2.6.

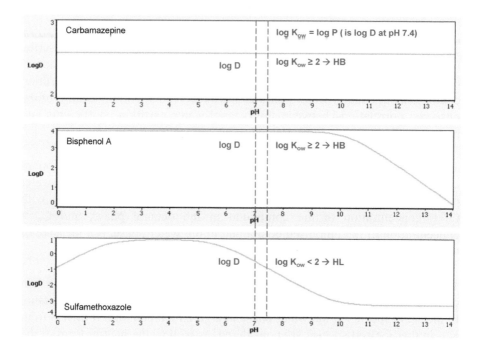

Fig. 2.6: Log D and log K$_{ow}$ for different organic compounds

Adapted from ADME/Tox Web Software

2.4   Characteristics of NF and RO membranes

2.4.1   Molecular weight cut-off

The molecular weight cut-off (MWCO) of a membrane is a parameter that indicates the relative size of membrane pores. The molecular weight cut-off corresponds to the molecular weight of a solute with a retention of 90%. MWCO is taken as a measure for the retention properties of the membrane. Thus, a mixture of similar uncharged molecules (e.g. dextrans, polyethylene glycol PEG) covering a range of molecular weights is used; retention is measured as a function of the molecular weight. The MWCO, sometimes called nominal molecular weight limit, is defined by its ability to retain a given percentage, e.g. 90% (Van der Bruggen et al., 1999) of a solute of a defined molecular weight. However, this definition is not explicit as the rejection percentage can vary between 60% and 90% (Bellona et al., 2004), and its measurement method has yet to be internationally standardized. Fig. 2.7 shows a schematic of diffused and sharp cut-off curves based on which membranes are sometimes characterised. The cut-off is often identified as a discontinuity; however, it will be difficult to identify a discontinuity even if the membrane contains a narrow distribution of pores (not dispersed) when concentration polarization occurs (Cardew and Le, 1998).

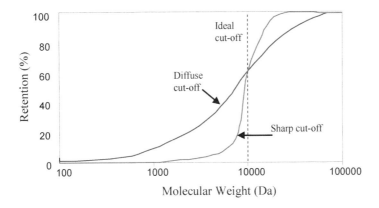

Fig. 2.7: Molecular weight cut-off curves of different membranes

Adapted from Cardew and Le, 1998

The MWCO, although being a useful parameter for evaluating the rejection of PhACs and EDCs, may not be relied on for a precise prediction of their rejection by NF/RO membranes. For instance, Kimura et al. (2004) observed the rejections of some neutral EDCs and PhACs to be less than 90% in spite of the fact that the molecular weights of those compounds were larger than the MWCO of the membrane inspected. Consequently, they suggested considering MWCO only for semi-quantitative prediction of organic micropollutant rejection by RO membranes.

2.4.2   Salt rejection

The desalting degree of a membrane is commonly described as the percent of salt rejection of a 500–2000 mg/L sodium chloride or magnesium sulphate solution at standard conditions specified by each membrane manufacturer for specific membrane types. Since the MWCO of a membrane is often manufacturer specific, salt rejection can be a useful parameter for comparisons between membranes. This is important because membranes with the same reported MWCO can have significantly different desalting degrees. The rejection of monovalent ions such as Na^+ and Cl^- by NF is in the range 50–90% or lower, while that by RO is over 99%. This difference is due to the more open pore structure of NF membranes. On the other hand,

NF membranes retain divalent ions such as Ca^{2+}, Mg^{2+}, CO_2^-, and SO_4^{-2} over 90%, while RO membranes retain them over 99% (Mulder, 1996).

2.4.3   Surface charge

Membranes in contact with an aqueous solution acquire an electric charge by various mechanisms: dissociation of surface functional groups, adsorption of ions from the solutions, and adsorption of polyelectrolytes, ionic surfactants and macromolecules (Elimelech et al., 1994). These charging mechanisms can take place on the exterior membrane surface and on the interior pore surface of the membrane because of the distribution of ions in solution to maintain the electroneutrality of the system (Schaep and Vandecasteele, 2001). The ion separation resulting from the electrostatic interactions between ions and membrane surface charge is based on the Donnan exclusion mechanism. In this mechanism the co-ions (which have the same charge as the membrane) are repulsed by the membrane surface and, to satisfy the electroneutrality condition, an equivalent number of counter-ions are retained which results in salt retention (Schaep et al., 1998; Childress and Elimelech, 2000).

Membrane surface charge is usually quantified by zeta potential measurements. Different studies have determined that pH had an effect upon the charge of a membrane due to the disassociation of functional groups (Childress and Elimelech, 2000; Xu and Lebrun, 1999). Zeta potentials for most membranes have been reported in many studies as becoming increasingly more negative as pH is increased and functional groups deprotonate (Braghetta et al., 1997; Lee et al., 2002). The surface charge of an NF membrane is negative, providing selective removal of charged contaminants (Bartels et al., 2008). Because many particles in water are also negatively charged, the negative surface charge enhances the removal of ionic compounds (Bellona and Drewes, 2007).

2.4.4   Hydrophobicity and hydrophilicity

A surface with a high affinity for water is called hydrophilic, while those with a low affinity are called hydrophobic (e.g. PTFE). The contact angle provides a measure of the hydrophilicity of the membrane surface (Cardew and Le, 1998). In this method, a drop of liquid is placed upon a flat (and smooth) surface and the contact angle is measured. For low affinity, the contact angle will have a value greater than 90°, whereas with high affinity the value will be less than 90° (Mulder, 1996). This is schematically shown in Fig. 2.8. Adsorption of organic compounds may be related to a change in hydrophobicity/ hydrophilicity of the membrane surface. Thus, a change in the contact angle may be a tool to measure adsorption. A significant increase in the contact angle for nanofiltration membranes by adsorption of natural organic matter has been reported (Roudman and DiGiano, 2000), which confirms the validity of this method.

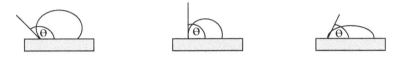

Fig. 2.8: Contact angle of liquid droplets on a solid material

Adapted from Mulder, 1996

2.5   Removal of PhACs and EDCs in water treatment

2.5.1   Conventional water treatment

Different studies have indicated that conventional water treatment with coagulation, sedimentation and filtration achieve low levels of removal for EDCs, PhACs, pesticides and herbicides (Vieno et al., 2006; Boyd et al., 2003; Adams et al., 2002).

Experiments to remove of selected pharmaceuticals (diclofenac, ibuprofen, bezafibrate, carbamazepine and sulphamethoxazole) by chemical coagulation in Milli-Q water, in lake water and in humic acid solutions using aluminium and ferric sulphate coagulants demonstrated that compounds, with the exception of diclofenac, were not removed by the coagulation processes in Milli-Q or lake water. Diclofenac was removed by up to 66% and 30% in pure water and in lake water, respectively, with ferric sulphate, but it was not removed with aluminium sulphate. Although conditions of high humic content, low coagulation pH and optimum ferric sulphate coagulant dose increased the removal of certain ionic pharmaceuticals (diclofenac, ibuprofen and bezafibrate), neutral pharmaceuticals were not removed at such conditions. The study concluded that coagulation does not entirely remove pharmaceuticals from water (Vieno et al., 2006)

Adams et al. (2002) concluded that little antibiotic removal (carbadox, sulfachlorpyridazine, sulfadimethoxine, sulfamerazine, sulfamethazine, sulfathiazole, and trimethoprim) resulted from coagulation, flocculation and sedimentation with alum or ferric salts, excess lime/soda ash softening, ultraviolet radiation, or ion exchange processes. The results of their study suggested that control of the studied antibiotics can be achieved at surface water treatment plants with common treatment steps, i.e. carbon sorption and oxidation with ozone or chlorine species. However, the same study concluded that reverse osmosis was effective for removal of the studied compounds with rejection levels greater than 90%.

2.5.2 Advanced water treatment

Based on the evidence that conventional water treatment presented limitations in removing emerging organic contaminants, researchers were motivated to look into other treatment alternatives, like activated carbon, oxidation, ozonation, ultraviolet radiation and membrane treatment.

Activated carbon

The addition of powdered activated carbon (PAC) or the use of granular activated carbon (GAC) beds during water treatment removes a number of organic contaminants by adsorption of the solutes within the pores of the carbon (Snyder et al., 2007). Nevertheless, initial water quality conditions can significantly affect the removal efficiency of GAC; for instance, the content of total organic carbon in the feed water will affect GAC performance for removing contaminants (Matsui et al., 2002a; Matsui et al., 2002b; Choi et al., 2005; Snyder et al., 2003); the removal efficiencies change drastically once the carbon bed reaches saturation, and subsequent solute breakthrough occurs. Removal percentages for many organic contaminants were moderate to excellent as demonstrated by the study by Westerhoff et al. (2005); they demonstrated that rejections of neutral organic solutes were in the range 44–99%, and that percentages of removals were dosage dependent.

Advanced oxidation

Advanced oxidation processes (AOP) are applied for transformation of emerging organic contaminants. Ozonation is an economically preferred alternative compared to ultraviolet light (Snyder et al., 2003). Ternes et al. (2002) demonstrated that ozonation showed removals of >99% for diclofenac and carbamazepine, however, clofibric acid was only partially removed (57%). Another study also demonstrated that ozone removed pharmaceuticals and atrazine (Hua et al., 2006). Removals have also been reported for ozone pilot plants with percentages between 25 and 83% (Snyder et al., 2008).

Membrane treatment

Membrane treatment options, particularly NF and RO, are preferred alternatives for the removal of emerging organic contaminants. In that sense, early studies (Van der Bruggen et al., 1998; Kiso et al., 2000 and 2001b) investigated the removal of micropollutants such as pesticides and organic solutes by NF membranes. These studies demonstrated that acceptable removals can be achieved for organic solutes and, furthermore, they proposed separation mechanisms such as size/steric exclusion, hydrophobic adsorption and electrostatic repulsion. In a full-scale membrane treatment plant, Drewes et al. (2002) determined the total removal of drugs after the RO treatment of a WWTP effluent previously treated using microfiltration. No pharmaceutical compounds were identified in the RO permeate. The study of Kimura et al. (2004) investigated the rejection of neutral (uncharged) EDCs and PhACs by RO membranes. The researchers performed experiments using RO membranes made of polyamide and cellulose acetate. One of their conclusions was that the polyamide membrane generally exhibited better rejection than the cellulose acetate membrane; however, the polyamide membrane did not exhibit complete rejection for the tested compounds.

2.5.3   Membrane rejection mechanisms

Steric hindrance

The rejection of uncharged trace organics by NF membranes is influenced by steric hindrance/size exclusion (Berg et al., 1997). Kiso et al. performed rejection experiments using hydrophobic compounds including aromatic pesticides, non-phenylic pesticides, and alkyl phthalates with NF membranes and concluded that compound rejection was correlated significantly with molecular width in addition to compound hydrophobicity (Kiso et al., 2000, 2001a, 2001b). One of their studies (2001a) showed that the rejection of hydrophilic solutes was controlled by molecular width rather than molecular weight. On the other hand, Ozaki and Li (2002) treated low molecular weight organic compounds (150 Da) using a low pressure RO membrane. They observed that rejection of uncharged organic compounds increased linearly with the molecular weight and molecular width. Kimura et al. (2003b) demonstrated through rejection experiments with disinfection by-products, EDCs, and PhACs that the rejection of uncharged compounds was influenced by their molecular size. However, their next study revealed that steric hindrance may not be the only factor to quantify the rejection of organic micropollutants (Kimura et al., 2004). From those studies, it should be noted that steric hindrance represented by MW and/or molecular width is one of the main factors affecting removal of uncharged organic contaminants.

Electrostatic repulsion

Electrostatic repulsion is known to be an important mechanism to separate charged solutes from membranes. Electrostatic repulsion is explained by the repulsive force between negatively charged compounds and negatively charged membrane surfaces. Numerous studies have shown electrostatic repulsion effects between charged compounds and membranes. Kimura et al. (2003b), for example, investigated the rejection of disinfectant by-products (DBPs), EDCs, and PhACs by NF and RO membranes as a function of their physicochemical properties and initial feed water concentrations. The results of their experiments indicated that negatively charged compounds could be rejected very effectively (i.e. >90%) regardless of other physicochemical properties of the tested compounds due to electrostatic exclusion. They also observed no time dependency for the rejection of charged compounds. However, rejection of uncharged compounds was generally lower (<90% except for one case) and was influenced mainly by the molecular size of the compounds. This rejection mechanism was also probed by Kim et al. (2005) who performed experiments with uncharged compounds under different pH conditions using negatively charged membranes. They observed good rejection of more polar/charged compounds where electrostatically hindered transport enhanced solute rejection.

Adsorptive interactions

The adsorption of hydrophobic compounds onto membranes may be an important factor in the rejection of micropollutants during membrane applications. Kiso et al. (2001a) showed over 99% compound rejection by NF membranes of more hydrophobic alkyl phthalates (log K_{ow} > 4.7) in their study. Surrogate compounds and three NF/RO membranes were studied by Kimura et al. (2003a) who found that the adsorption of hydrophobic compounds was significant for neutral compounds and ionisable compounds with an electrostatically imposed neutral presence. Considering their operating conditions, the permeate flow rate (flux) had a significant effect on the degree of compound adsorption. Adsorption results observed in dynamic filtration tests with those in static batch adsorption tests suggested that more adsorption sites were accessible for molecules during membrane filtration due to the pressurized advective flow. Kimura et al. (Ibid) also observed that the concentration of the tested compounds changed during filtration tests due to adsorption. Therefore, an accurate evaluation of a given membrane in terms of the rejection of a hydrophobic compound is not possible until saturation of the membrane with the compound of interest is accomplished;

this means that a relatively large amount of feed water must be filtered to reach saturation conditions and to avoid an overestimation of rejection.

In another study, adsorption of neutral, moderately hydrophobic compounds (bromoform and chloroform) by a more hydrophobic membrane was observed when compared to a less hydrophobic membrane (Xu et al., 2005). Schäfer et al. (2003) reported that estrone can adsorb (partition) onto the membrane to some extent and concluded that both size exclusion and adsorption are essential to maintain high initial retention by NF membranes. The removal mechanisms of four natural steroid hormones: estradiol, estrone, testosterone and progesterone by two NF membranes with different permeabilities and salt retention characteristics were studied in an investigation carried out by Nghiem et al. (2004). The results indicated that, at the early stages of filtration, adsorption (or partitioning) of hormones to the membrane polymer was the dominant removal mechanism. The final retention stabilises when the adsorption of hormones into the membrane polymer has reached equilibrium because of the limited adsorptive capacity of the membrane. The overall hormone retention was lower than that expected based solely on the size exclusion mechanism. That behaviour was attributed to partitioning and the subsequent diffusion of hormone molecules in the membrane polymeric phase, which ultimately resulted in a lower retention. However, their size exclusion model used underestimated NF pore radii of membranes NF-90 (0.34nm) and NF-270 (0.42nm).

The study by Majewska-Nowak et al. (2002) found that pesticides such as atrazine could adsorb to organic matter present in feed water, increasing rejection as a result of increased size and the electrostatic interaction between the organic and the membrane. Furthermore, Comerton et al. (2008) observed higher rejection of EDCs and PhACs from natural waters than from Milli-Q water, likely because of membrane fouling and compound interactions with the water matrix itself. While comparing rejection of those compounds from natural waters (filtered and raw lake water and membrane bio-reactor (MBR) effluent), they concluded that the presence of calcium ion lowered the rejection by interfering with compound-organic matter complex formations. By contrast, a different study indicated that the feed water matrix had almost no effect on total pesticide rejection by RO membranes, although some few variations were observed for individual pesticides and specific membranes (Taylor et al., 2000). In the same study, source water matrix tests also confirmed that total pesticides rejection was not affected by different natural organic matter compositions present in the feed water. In summary, it is unclear if feed water types, fouled membranes and clean membranes may possibly result in improved, diminished or almost equal removal of organic contaminants.

In summary, different studies have demonstrated that nanofiltration and reverse osmosis membranes are capable of removing low concentrations of organic solutes present in water samples. These studies have also demonstrated that physicochemical properties of solutes and membrane characteristics may explain transport, adsorption and removal of neutral organic compounds by different solute-membrane mechanisms such as size/steric exclusion, hydrophobic adsorption and partitioning (Van der Bruggen et al., 1999; Ozaki and Li, 2002; Van der Bruggen and Vandecasteele, 2002; Schafer et al., 2003; Kimura et al., 2003b and 2004; Nghiem et al., 2004; Bellona and Drewes, 2005; Xu et al., 2005).

2.6   Multivariate data analysis

2.6.1   Multiple linear regression

Multiple linear regression (MLR) is a method of analysis for assessing the strength of the relationship between a set of explanatory variables known as independent variables and a single response or dependent variable. Applying multiple regression analysis to a set of data results in what are known as regression coefficients, one for each explanatory variable. The multiple regression model for a response variable, y, with observed values, $y_1, y_2, ..., y_n$ (where n is the sample size) and q explanatory variables, $x_1, x_2, ..., x_q$ with observed values, $x_{1i}, x_{2i}, ..., x_{qi}$ for $i = 1, ..., n$, is

$$y_i = \beta_o + \beta_1 x_{1i} + \beta_2 x_{2i} + \cdots + \beta_q x_{qi} + \varepsilon_i \qquad (2.3)$$

The regression coefficients, $\beta_0, \beta_1, ..., \beta_q$, are generally estimated by least squares. The term ε_i is the residual or error for individual i and represents the deviation of the observed value of the response for this individual from that expected by the model. These error terms are assumed to have a normal distribution with variance σ^2. The fit of a multiple regression model can be judged by calculating the multiple correlation coefficient R (also referred to as regression coefficient), defined as the correlation between the observed values of the response variable and the values predicted by the model. The squared value of R (R^2) gives the proportion of the variability of the response variable accounted for by the explanatory variables. Analysis of variance (ANOVA) will provide an *F-test* of the null hypothesis that each of $\beta_0, \beta_1, ..., \beta_q$, is equal to zero, or in other words that R^2 is zero (Landau and Everitt, 2004).

2.6.2   Principal component analysis

Principal component analysis (PCA) is a method that allows the simplification of many variables into a group of a few variables that might be measuring the same principles of a system. It may occur that a system considers an abundance of variables to explain a process; in this case principal component analysis reduces the redundancy of information. The general objectives of PCA are data reduction and interpretation. Although p components (variables) are required to reproduce the total system variability, often much of this variability can be accounted for by a small number k of principal components. Thus, there is as much information in the k components as there is in the original p variables. Comprehensive details about the theory of PCA are found elsewhere (Everitt and Dunn, 2001, Jolliffe, 2002; Johnson and Wichern, 2007). PCA is a method of reduction that aims to produce a small number of derived variables that can be used in place of the larger number of original variables, while retaining as much as possible of the variation present in the data set. A further objective of PCA is to simplify subsequent analysis of the data. A summary of the theory of PCA has been adapted from multivariate analysis and statistics books (Everitt and Dunn, 2001; Jolliffe, 2002; Johnson and Wichern, 2007). The summary is presented in Appendix A.

2.6.3   Partial least-squares regression

Partial least-squares (PLS) regression finds components from $\mathbf{X}$ that are also relevant for $\mathbf{Y}$. Specifically, PLS regression searches for a set of components (called latent vectors) that performs a simultaneous decomposition of $\mathbf{X}$ and $\mathbf{Y}$ with the constraint that these components explain as much as possible about the covariance between $\mathbf{X}$ and $\mathbf{Y}$. This step generalises PCA. The goal of PLS regression is to predict $\mathbf{Y}$ from $\mathbf{X}$ and to describe their common structure (Abdi, 2003; Jørgensen and Goegebeur, 2006). Principal component regression (PCR) is a method in which the components from the principal component method are used for regression. Hence, the principal components of the matrix $\mathbf{X}$ are used as regressors of a dependent $\mathbf{Y}$. The orthogonality of the principal components eliminates the multi co-linearity problem. But, the problem of choosing an optimum subset of predictors remains. A possible strategy is to keep only a few of the first components. But, they are chosen to explain $\mathbf{X}$ rather than $\mathbf{Y}$, and, therefore, nothing guarantees that the principal components, which "explain" $\mathbf{X}$, are relevant for $\mathbf{Y}$. Problems may arise, however, if there is a lot of variation in $\mathbf{X}$. PCR finds, somewhat uncritically, those latent variables that describe as much as possible of the variation in $\mathbf{X}$. But sometimes the variable itself gives rise to only small variations in $\mathbf{X}$, and if the interferences vary a lot, then the latent variables found by PCR may not be particularly good at describing $\mathbf{Y}$. In the worst case, important information may be hidden in directions in the **X**-space that PCR interprets as disturbance and, therefore, leaves out. PLS regression is able to cope better with this problem by forming variables that are relevant for describing **Y** (Abdi, 2003; Jørgensen and Goegebeur, 2006).

2.7   **Artificial neural networks**

An introduction to artificial neural networks is presented in Appendix B.

Chapter 3

Removal of PhACs and EDCs by clean NF membranes and NF membranes fouled with sodium alginate

Based on parts of:
- Rejection of Organic Micropollutants by Clean and Fouled Nanofiltration Membranes, AWWA Membrane Technology Conference & Exhibition, March 15-18, 2009, Memphis, Tennessee.
- Rejection of Pharmaceutically Active Compounds and Endocrine Disrupting Compounds by Clean and Fouled Nanofiltration Membranes. Water Research, 43 (2009), 2349-2362.

3.1   Introduction

In the future, the ever-growing demand for drinking water will lead many cities to implement indirect water reuse programs, where wastewater effluent is used to augment their drinking water sources. Pollution of those sources with organic micropollutants such as endocrine disrupting compounds (EDCs) and pharmaceutically active compounds (PhACs) which have been detected in water supplies and wastewater effluents around the world pose negative health effects for consumers and the environment (Kolpin et al., 2002; Snyder et al., 2003; Ternes et al., 2004). Membrane filtration technology, particularly nanofiltration (NF) and reverse osmosis (RO), has demonstrated promising results with the rejection of PhACs and EDCs. In order to examine the ability of RO membranes to retain PhACs and EDCs, Kimura et al. (2004) showed that the polyamide membranes exhibited better rejection than cellulose acetate membranes. Results from other investigations showed that, due to electrostatic repulsion, the rejection of negatively charged compounds was effective and varied from 89% to over 95% by NF membranes and exceeded 95% by ULPRO and RO membranes (Kimura et al., 2003b; Xu et al., 2005; Nghiem et al., 2006).

The results from a study on removal of hormones and pharmaceuticals from treated sewage indicated that ozonation, microfiltration and nanofiltration were partially effective whereas RO treatment was the most successful in the removal of target residuals (Khan et al., 2004). Ozaki and Li (2002) showed that the rejection of organic compounds by ultra-low pressure RO (ULPRO) increased linearly with the molecular weight and molecular width, while investigating the rejection of DBPs, EDCs and PhACs by polyamide NF/RO membranes. Fouling may alter membrane surface characteristics in terms of the contact angle, zeta potential, functionality and surface morphology, which potentially affect transport of contaminants compared to non-fouled membranes; for instance Ng and Elimelech (2004) observed a decline in the rejection of hormones by RO membranes after colloidal fouling. Furthermore, findings of another study indicated that membrane fouling significantly affected the rejection of organic micropollutants by cellulose acetate RO, NF and ULPRO membranes (Xu et al., 2006). After the organic fouling of membranes, Agenson and Urase (2007) observed a decrease in rejection of high molecular weight (MW) neutral organics by NF/RO membranes; however, rejection of low MW compounds was reported to have increased. In addition, Makdissy et al. (2007) observed lower rejection of EDCs and personal care products (PCPs) by NF membranes fouled by surface water than by clean membranes. As many of these studies illustrate, membrane fouling has the potential to affect rejection mechanisms of organic solutes as a result of modified electrostatic, steric and hydrophobic/hydrophilic solute-membrane interactions. However, reported results are complicated to follow due to the variety of foulants and particular interactions with each membrane type and feed water composition that leads to a diversity of explanations for observed rejections. This investigation attempts to overcome that diversity using defined groups of organic compounds, well-characterised polyamide nanofiltration membranes and a surrogate foulant. An emphasis on the interaction between clean membranes and compounds is given in this chapter, considering rejections after steady-state saturation of the membrane and adsorption on the membranes. In this chapter, there is a description of the use of sodium alginate as the foulant, a hydrophilic (anionic) polysaccharide, which forms a uniform film on the membrane surface altering the electric and hydrophobic membrane characteristics. In the next chapter, there is an explanation of the differences between the rejection of clean and fouled NF membranes using an additional foulant surrogate (dextran) and natural organic matter (NOM) from surface water.

3.2   Theory

3.2.1   Hydrodynamic conditions

The back-diffusion transport in the boundary layer of a membrane is defined by the mass transfer coefficient, k, a function of the diffusion coefficient, feed channel hydrodynamics and the cross-flow velocity. The water permeation flux, J (cm/s), can be compared with the k value to determine J/k, which indicates the ratio of the initial transport of the compound/molecule to the membrane surface by convection to its back transport by diffusion. A J_0/k ratio (where J_0 is the initial pure water permeation flux) can be used to control hydrodynamic operating conditions while simulating the differences in membrane permeability and thereby facilitating the comparison of solute rejection and flux decline by different membranes (Cho et al., 2000). J and k can be calculated by the following equations for a thin-channel-type module (Porter, 1972):

$$J = \frac{Q_p}{A_m} \tag{3.1}$$

$$U = \frac{Q_c}{A_{cross}} \tag{3.2}$$

$$k = 1.62 \left(\frac{UD^2}{d_h L} \right)^{0.33}$$
(3.3)

where Q_p and A_m are water permeate flow rate (cm³/s) and membrane surface area (cm²), respectively. U is the average velocity of the feed fluid (cross-flow velocity, cm/s), Q_c is the concentrate flow rate (cm³/s), A_{cross} is the cross-sectional area of the channel (cm²), D is the diffusion coefficient of the solute in water (cm²/s) and is estimated by the Hayduk and Laudie method with Eq. 3.4 (Schwarzenbach et al., 2003), d_h is the equivalent hydraulic diameter (cm), and L is the channel length (cm).

$$D = \frac{13.26 \times 10^{-5}}{\eta^{1.14} (MV)^{0.589}}$$
(3.4)

where η is the viscosity of water in centipoise (10^{-2} g/cm-s) and MV is the molar volume in cm³/mol. Calculation of MV is explained in Section 3.2.2.

In NF and RO, solutes present in the feed are convected to the membrane with water. Concentration polarization is defined as the accumulation of solutes close to the membrane. The balance between convection towards the membrane, due to water flux (J), and back transport from the membrane to the bulk solution due to the concentration gradient determines the magnitude of concentration polarization (Schaffer et al., 2005). The concentration polarization equation derived by Brian (1966) for the film model is obtained by a boundary-layer mass balance where net convection equals back diffusion,

$$\frac{c_m - c_p}{c_b - c_p} = \exp(J/k)$$
(3.5)

In this equation, c_m is the concentration in the feed solution at the membrane surface, c_b is the concentration of the feed solution (bulk), c_p is the permeate concentration, J is the flux and k is the mass transfer coefficient. The ratio J/k is also known as the Peclet number. The mass transfer coefficient is equal to D/δ, where δ is the thickness of the boundary layer. Wijmans et al. (1996) present an alternative form of Eq. 3.5; their equation is

$$\frac{1/E_0 - 1}{1/E - 1} = \exp(J/k)$$
(3.6)

where E is defined as c_p/c_b, and E_0 is c_p/c_m.

The concentration polarization modulus is defined as the ratio c_m/c_b and measures the extent of concentration polarization. The ratio is equal to E/E_0 and from Eqs 3.5 and 3.6 can be represented as

$$\frac{c_m}{c_b} = \frac{\exp(J/k)}{1 + E_0\left[\exp(J/k) - 1\right]} \qquad (3.7)$$

3.2.2   Calculation of physicochemical properties

Size descriptors include molar volume (MV), molecular length, molecular width, molecular depth and equivalent molecular width. The molecular length is defined as the distance between the two most distant atoms of a particular molecule. The molecular width and molecular depth (width > depth) are measured by projecting the molecule on a plane perpendicular to the length axis. The equivalent molecular width is defined as the geometric mean of width and depth (Santos et al., 2006). The molecular dipole moment is equal to the vector sum of the individual bond dipole moments. The octanol-water partition coefficient (log K_{ow}), often used to describe hydrophobicity, is a measure of the equilibrium concentration of a compound between octanol and water. The ratio of the equilibrium concentrations of all species (ionised and unionised) of a particular molecule in the octanol phase and in the water phase is expressed as log D; it differs from log K_{ow} in that ionised species as well as the neutral form of the molecule are considered. Values of log D were calculated by ADME/Tox web software. Values of log K_{ow} were obtained from SRC Physprop experimental database. The dipole moment was calculated by Chem3D Ultra 7.0 software (Cambridgesoft). Compound molar volume was calculated using the program ACD/ChemSketch Properties Batch (ACD/Labs). Molecular Modeling Pro (ChemSW) was used to compute molecular length and equivalent width.

3.2.3   Rejection types

When experiments are conducted in the recycle mode in which permeate and concentrate are recirculated into the feed tank, two rejections may be observed. R_1 is defined as rejection under "steady-state" conditions and is a measurement of membrane rejection after saturating (pre-equilibration of) the membranes with solutes. R_2 is defined as rejection that incorporates adsorption of solutes onto the membrane surface. Compound rejections (R_1 and R_2) have been determined using Eq. 3.8 and Eq. 3.9:

$$R_1(\%) = \left(1 - \frac{C_p}{C_f}\right) \times 100 \tag{3.8}$$

$$R_2(\%) = \left(1 - \frac{C_{p0}}{C_{f0}}\right) \times 100 \tag{3.9}$$

where C_p is the permeate concentration after a saturation time, C_{p0} is the initial permeate concentration, C_f is the feed concentration after saturation and C_{f0} is the initial feed concentration. It can be assumed that compound losses on non-membrane components are negligible when the membrane cells, the tubing and the reservoir are made of stainless steel; volatilization can also be disregarded when compounds under investigation have very low Henry's law constants. It was mentioned that R_2 is a rejection that consider adsorption. Adsorption mainly occurs due to the hydrophobicity of the compound that produces partitioning of the compound onto/into the membrane surface.

3.2.4 Fouling protocol with sodium alginate

The extent of membrane fouling was described by permeate flux decline (J/J_0) as a function of the amount of dissolved organic matter (DOC) delivered to the membrane per unit surface area (cm²). To estimate the amount of DOC delivered to the membranes (DOC_m), a mass balance was carried out for the experimental system,

$$DOC_m = V_i C_{DOCi} - (V_f \times C_{DOCf} + V_{cp} \times C_{DOCcp}) \tag{3.10}$$

where, V_i and V_f are initial feed volume and remaining feed volume after time t respectively, V_{cp} is the volume of the combined concentrate and permeate volumes (in L), C_{DOCi} is the DOC concentration in the feed at the beginning, C_{DOCf} is the DOC concentration in the feed after time t, and C_{DOCcp} is the DOC concentration in the mixture of concentrate and permeate (in mg/L). In order to demonstrate no sorption of DOC to the apparatus, a mass balance of a recirculation experiment with DOC without using a membrane was performed; negligible sorption was observed (<1%). Sodium alginate was used as a surrogate of polysaccharides; the study of Lee et al. (2004) concluded that polysaccharides were important membrane foulants.

Alginate is frequently used as a model for organic matter of algae origin (Henderson et al., 2008).

3.3   Experimental

3.3.1   Chemicals and membranes

The PhACs (caffeine, sulphamethoxazole, acetaminophen, phenacetin, phenazone, carbamazepine, naproxen, ibuprofen, metronidazole), EDCs (17β-estradiol, estrone, bisphenol A, nonylphenol, atrazine) and sodium alginate were purchased from Sigma-Aldrich (Schnelldorf, Germany). Potassium chloride, sodium hydroxide, hydrochloric acid and magnesium sulphate anhydrous were purchased from J.T. Baker (Deventer, the Netherlands). Sodium bisulfite was obtained from Acros Organics (Geel, Belgium). Two thin-film composite NF membranes were selected for this study: NF-200 and NF-90 (Dow-Filmtec, Dow Chemical Co., Midland, MI) made of aromatic polyamide. The selection of membranes was based on a qualitative rejection assessment of emerging contaminants with a molecular weight of more than 150g/mol by membranes with a MWCO between 200 and 300Da.

3.3.2   Experimental setup and experimental conditions

The membranes were used as flat-sheet specimens in two-parallel cross-flow filtration units. Each cell provided an effective membrane area of 139 cm^2. Spacers and shims (to improve the hydrodynamic conditions and to control the height of the channel) were used in the experiments. The experimental setup consisted of two filtration SEPA CF II (GE Osmonics, Minnetonka, MN) cells and cell holders in parallel, in order to increase permeate production and achieve the desired hydrodynamic conditions, two hydraulic pumps (Power Team, Bega Int. BV, the Netherlands), a 60 L stainless steel tank (Tummers, Netherlands), a positive displacement pump (Hydra-Cell pump, Wanner Eng. Inc., Minneapolis, MN), a frequency converter (VLT microdrive, Danfoss, SA), a chiller/heater (Julabo, Germany), control needle valves, pressure gauges, flow meters, a proportional pressure relief valve and stainless steel tubings (Swagelok BV, the Netherlands), a digital balance (Sartorius, Germany) and a computer for flow rate data acquisition. A scheme of the experimental setup is shown in Fig. 3.1. A required volume of KCl solution was added to maintain an ionic strength of 10mM KCl. Prior to and after starting a filtration test, the pH of the solution was adjusted to 7 by adding 0.1M NaOH as needed. Filtration experiments were performed for a total of 96 hours in order to provide pre-equilibration of the membrane to avoid overestimating rejection (Kimura et al., 2003a). Membranes were cleaned and compacted by filtering demineralized water for six hours. The experiments were conducted in the recycling mode in which permeate and concentrate were recirculated into the feed tank for 72 hours; then, within the last 24 hours, permeate was collected. The feed solution for all the experiments contained a cocktail of 14 compounds (concentration ranging from 6.5 to 13µg/L, except for nonylphenol with a concentration of 65µg/L); the intended concentration was 10µg/L per compound, but pipet measuring inaccuracy from stock solutions, weighing innacuracy and differences in solubilities resulted in variations in concentrations (from 10µg/L) as corroborated by the results of the chemical analyses. All of the experiments were carried out at a controlled temperature of 20°C ±0.5. A specific flux decline of 15% of the initial flux (a realistic condition in membrane operation) was targeted for the membrane fouling protocol using a feed solution containing ~10mg/L DOC of sodium alginate. This high concentration (compared to real waters) was used to accelerate the formation of the fouling layer. The model foulant sodium alginate is a hydrophilic anionic polysaccharide that is produced by algae and bacteria. It has a high molecular weight (20,000 to 2,000,000 g/mol) and is a copolymer of two monomers; β-D-mannuronic acid and α-L-guluronic acid residues. The ratio of the monomers and the structure of the polymer vary widely (Van de Ven et al., 2008).

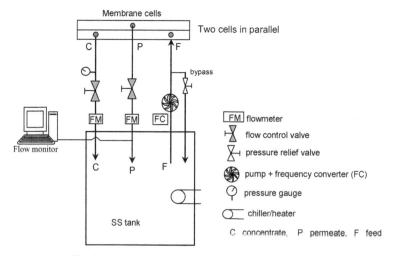

Fig. 3.1: Layout of experimental setup

The 96-hour duration of experiments in this study provided adequate membrane saturation for the recirculation of 50L of feed solution. Kimura et al. (2003a), in an attempt to establish an experimental protocol for filtration of some hydrophobic compounds, demonstrated that a "quasi-saturation" of the tested membrane was reached after about 20 hours of operation using a feed solution of 100ppb concentration. They further suggested that for low concentration feed, the filtration time should be extended and a large volume of feed should be circulated in order to achieve sufficient membrane saturation. It is therefore assumed that the 96-hour duration in this study facilitated adequate membrane saturation upon recirculation of 50L of feed containing 14 compounds. In order to investigate the interaction between different compounds that may lead to biased results, a bi-solute control experiment was performed with a clean NF-90 membrane. In the control experiment, the solution contained ibuprofen, a hydrophobic ionic compound, and estrone, a neutral compound. Rejection of ibuprofen and estrone in the bi-solute control experiment were 97 and 88%, respectively. These rejections were slightly less than in a cocktail experiment (ibuprofen 99% and estrone 92%). The cocktail experiment contained as feed 14 compounds shown in Table 3.1.

Table 3.1: List of compounds and physicochemical properties

Name	Name ID	Molec. weight (g/mol)	log K_{ow}[a]	logD[b] (pH 7)	Dipole moment (debye)[c]	Molar volume[d] (cm^3/mol)	Molec. length (nm)[d]	Equiv. width (nm)[d]	Classification[e]
Acetaminophen	ACT	151	0.46	0.23	4.55	120.90	1.14	0.53	HL-neutral
Phenacetine	PHN	179	1.58	1.68	4.05	163.00	1.35	0.54	HL-neutral
Caffeine	CFN	194	-0.07	-0.45	3.71	133.30	0.98	0.70	HL-neutral
Metronidazole	MTR	171	-0.02	-0.27	6.30	117.80	0.93	0.66	HL-neutral
Phenazone	PHZ	188	0.38	0.54	4.44	162.70	1.17	0.66	HL-neutral
Sulphamethoxazole	SFM	253	0.89	-0.45	7.34	173.10	1.33	0.64	HL-ionic
Naproxen	NPN	230	3.18	0.34	2.55	192.20	1.37	0.76	HB-ionic
Ibuprofen	IBF	206	3.97	0.77	4.95	200.30	1.39	0.64	HB-ionic
Carbamazepine	CBM	236	2.45	2.58	3.66	186.50	1.20	0.73	HB-neutral
Atrazine	ATZ	216	2.61	2.52	3.43	160.07	1.26	0.74	HB-neutral
17 β-estradiol	E2	272	4.01	3.94	1.56	232.60	1.39	0.74	HB-neutral
Estrone	E1	270	3.13	3.46	3.45	232.10	1.39	0.76	HB-neutral
Bisphenol A	BPA	228	3.32	3.86	2.13	199.50	1.25	0.79	HB-neutral
Nonylphenol	NPL	220	5.71	5.88	1.02	236.20	1.79	0.66	HB-neutral

[a] Experimental database: SRC PhysProp Database
[b] ADME/Tox Web Software
[c] Chem3D Ultra 7.0
[d] Molecular Modeling Pro
[e] HL = Hydrophilic, HB = Hydrophobic, Hydrophobic if log K_{ow} >2

3.3.3   Analyses of compounds and analytical equipment

The water samples containing PhACs and EDCs (with the exception of atrazine) were analysed by Technologiezentrum Wasser TZW (Karlsruhe, Germany). The detection limit was 10ng/L per compound. The uncertainty of estimates was ±15% according to a validation method of the analysis protocol. Recoveries were between 70–100%.

For group A (acetaminophen, caffeine, carbamazepine, ibuprofen, naproxen, phenacetine and phenazone), the analysis was conducted by an HPLC-MS-MS method. Prior to the analysis, an automated solid-phase extraction on plastic cartridges filled with 200mg of Bakerbond SDB 1 material (Mallinckrodt Baker, Deventer, the Netherlands) was performed. Subsequently, a determination of the analyte was carried out by injecting it two-fold into an HPLC-ESI-MS-MS system. More information about the analytical protocol has been previously published (Sacher et al., 2008). For group B (metronidazole and sulphamethoxazole), the analysis was conducted by an HPLC-MS-MS method. Prior to the analysis, an automated solid-phase extraction on plastic cartridges filled with 0.1g of Isolut ENV+ material (Separtis, Grenzach-Wyhlen, Germany) was performed. Determination of the analyte was carried out by HPLC-ESI-MS-MS. More details about the method can be found elsewhere (Sacher et al., 2001). For group C (17β-estradiol, estrone, bisphenol A and nonylphenol), the analysis was conducted by gas chromatography / mass spectrometry (GC/MS). Prior to the determination, an automated solid-phase extraction on plastic cartridges filled with 200mg of bondelut material (Fa. Varian, Darmstadt, Germany) was conducted. Determination of this analyte was conducted by GC/MS using a 6890 GC/MS system (Aglient Technologies, Waldbronn, Germany). More details of this method were described in a previous publication (Schlett and Pfeifer, 1996). Concentrations of atrazine were determined using microplate ELISA kits (Abraxis, Norway).

The pH of the solutions was measured using a calibrated Metrohm 691 pH-meter (Metrohm AG, Herisau, Switzerland); the electrical conductivity and temperature were measured with a WTW Cond 330i (WTW GmbH, Weilheim, Germany) portable conductivity meter. DOC was analysed using a Shimadzu TOC - VCPN total organic carbon analyzer (Shimadzu, Japan).

To determine membrane hydrophobicity, the contact angles of clean and fouled membrane surfaces were measured with a CAM200 optical contact angle and surface tension meter (KSV Instruments, Finland) at Delft University of Technology. To measure the contact angles, the sessile drop method was used. The membrane samples were dried for 24 hours at room temperature. Precautions were taken to avoid alteration of sample surfaces, and at least five readings were taken at different positions across the sample. The surface charge, in terms of zeta potential, of the clean and fouled membranes was quantified using ELS-8000 zeta potential analyzer (Otsuka Electronics, Japan). The zeta potential analyses were determined using a Milli-Q water solution at pH 7 and ionic strength of 10mM KCl. The zeta potential was determined using the electrophoresis method using a cell consisting of a membrane specimen and quartz cells. The zeta potential was calculated from the electrophoretic mobility using the Smoluchowski formula, a detailed explanation of this calculation was provided in a previous publication (Shim et al., 2002).

3.3.4   Classification of compounds

Based on dissociation species at pH 7 and log K_{ow} values, the compounds were classified as hydrophilic neutral, hydrophilic ionic, hydrophobic ionic and hydrophobic neutral. Compounds were classified as ionic or neutral based on dissociated species at pH 7. Compounds with log $K_{ow} \geq 2$ were referred to as hydrophobic compounds; those with log $K_{ow} < 2$ were classified as hydrophilic. This classification was based on an early reference (Connell, 1990). However, a log K_{ow} (log D) higher or equal than 3 can also be used to refer an organic compound as hydrophobic. Table 3.1 shows the calculated values of molecular weight, log K_{ow} and log D at pH 7, dipole moment, molar volume, molecular length and equivalent width.

3.4   Results and discussion

3.4.1   Concentration polarization

In order to evaluate rejection by both membranes, two average J/k ratios were studied: 0.5 and 0.8, which also corresponded to recoveries of 3% and 8%, respectively. Cross-flow velocities were in the range of 3 to 7.6 cm/s. Other hydrodynamic parameters are shown in Table 3.2.

The concentration polarization (CP) modulus, previously defined in Section 3.2.1 by Eq. 3.7, was calculated after solving the equation for c_m and E_0 (c_p/c_m). According to Table 3.3, the concentration polarization modulus for the flat sheet cross-flow NF-200 (NF-90) experiments considering organic solutes was 1.4 (1.6) for J/k – 0.5 and recovery 3%, and 1.7 (2.1) for J/k = 0.8 and recovery 8%. The main consideration for calculation of an average concentration polarization was the assumption of a mean diffusion coefficient in water (D) with its respective mean back diffusion mass transfer coefficient (k) for the group of organic solutes used in the experiments. The biggest compound (min D for 17β-estradiol) showed concentration polarisation of 1.5 (1.9) for NF-200, J/k = 0.5 and recovery 3% (J/k = 0.8 and recovery 8%). The smallest compound (max D for metronidazole) showed concentration polarisation of 1.2 (1.3) for NF-200, J/k = 0.5 and recovery 3% (J/k = 0.8 and recovery 8%); observe that average concentrations of organic solutes on the membrane were in the range of 10.2–17.4µg/L.

The concentration polarization for NaCl was 1.2, and that for MgSO$_4$ was 1.4, both for NF-200 in flat sheet cross-flow configuration at J/k = 0.5 and recovery 3% (Table 3.3).

The CP modulus has also been calculated for clean NF membrane elements (8×40"). The calculations were made for membrane strips in an envelop configuration. The assumptions and results are presented in Appendix C. Concentration polarization was calculated for organic solutes (mean D and k), NaCl and MgSO$_4$. The CP modulus for NF-200 (NF-90) was 1.6 (1.9) at J/k = 0.7 and recovery 8%, for organic solutes. Sodium chloride resulted in a modulus of 1.3 (1.4) for NF-200 (NF-90) at J/k = 0.7 and recovery 8%. The modulus for magnesium sulphate was 1.5 (1.5) for NF-200 (NF-90) at J/k = 0.7 and recovery 8%.

Table 3.2: Configuration parameters and hydrodynamic conditions in lab-scale unit (flat sheet membranes)

Parameter	Unit	NF-200		NF-90	
Membrane area (A$_m$)	cm²	278	278	278	278
Cross-section area (A$_{cross}$)	cm²	0.36	0.36	0.36	0.36
Pressure	kPa	483	483	276	276
Cross flow velocity (U)	cm/s	6.8	3.0	7.6	3.0
Mean diffusion coeff. (D)	cm²/s	6.30E-06	6.30E-06	6.30E-06	6.30E-06
Equiv. hydraulic diameter	cm	0.075	0.075	0.075	0.075
Channel length (L)	cm	14.6	14.6	14.6	14.6
Mean back diffusion mass transf. coef. (k)	cm/s	1.1E-03	8.3E-04	1.1E-03	8.3E-04
Permeate flow (Q$_p$)	mL/min	9	11	10	11
Concentrate flow, (Q$_c$)	mL/min	291	127	323	127
Flux, J = Q$_p$/A$_m$	L/m²-h	19.4	23.7	21.6	23.7
	cm/s	5.4E-04	6.6E-04	6.0E-04	6.6E-04
J/k		0.5	0.8	0.5	0.8
Recovery	%	3	8	3	8

Table 3.3: Calculation of concentration polarization in a lab-scale unit

Parameter	Unit	NF-200 $J/k=0.5$ Rec.=3%	NF-200 $J/k=0.8$ Rec.=8%	NF-90 $J/k=0.5$ Rec.=8%	NF-90 $J/k=0.8$ Rec.=8%
CP average org. sol.		1.4	1.7	1.6	2.1
c_m average org. sol.	μg/L	10.2	12.7	12.7	17.4
$E_0 = c_p/c_m$ average		0.294	0.236	0.063	0.034
k (min D)		9.8E-04	7.5E-04	1.0E-03	7.5E-04
k (max D)		1.3E-03	9.7E-04	1.3E-03	9.7E-04
CP (min D)		1.5	1.9	1.7	2.3
c_m (min D)	μg/L	5.6	7.1	7.6	7.8
$E_0 = c_p/c_m$ (min D)		0.252	0.212	0.052	0.051
CP (max D)		1.2	1.3	1.5	1.9
c_m (max D)	μg/L	13.7	14.7	17.7	22.3
$E_0 = c_p/c_m$ (max D)		0.422	0.490	0.119	0.063
k (NaCl)		2.2E-03	1.6E-03	2.2E-03	1.6E-03
k (MgSO$_4$)		1.6E-03	1.2E-03	1.7E-03	1.2E-03
CP (NaCl)		1.2	1.3	1.3	1.4
c_m (NaCl)	μg/L	2.4E+06	2.8E+06	2.7E+06	3.2E+06
$E_0 = c_p/c_m$ (NaCl)		0.273	0.234	0.086	0.072
CP (MgSO$_4$)		1.4	1.7	1.4	1.7
c_m (MgSO$_4$)	μg/L	2.9E+06	3.7E+06	3.0E+06	3.7E+06
$E_0 = c_p/c_m$ (MgSO$_4$)		0.019	0.015	0.011	0.009

The CP modulus (for organic solutes) was 1.6 in experiments with flat-sheet membranes at lab-scale, and was equal to the CP modulus in NF-90 membrane elements at recovery 8% (see Appendix C). The CP modulus (for organic solutes) was slightly higher in experiments with flat-sheet membranes at lab-scale (1.7) compared to NF-200 membrane elements with a CP modulus of 1.6 at recovery 8%. The concentration polarization increases in the element at higher recoveries and vary with the distance to the center of the spiral wound element. At 15% recovery the modulus was 1.7 (NF-200, Appendix C). Therefore, the calculations suggest that, in terms of concentration polarization, the experiments performed with flat-sheet membranes at lab-scale may be comparable to membrane elements (8×40").

Baker (2004) reports that reverse osmosis for seawater desalination will present a concentration polarization modulus of 1.3, and it will increase to 1.5 for brackish water desalination. Baker's calculation corresponds with the calculations for sodium chloride and magnesium sulphate performed in Appendix C. Ma et al. (2004) also calculated concentration polarization in spiral wound RO modules by using finite element models and found values of more than 1.2 (depending on the profiles) for sodium chloride solutions in feed channel with spacers. Moreover, the calculations of this chapter are also comparable with the calculations of CP (known as beta factor) obtained with the software IMS Design (Hydranautics) for sodium chloride (see Appendix C).

Figure 3.2 illustrates the response of the concentration polarization modulus (c_m/c_b) to the Peclet number (J/k) and the factor E_0 (c_p/c_m), the figure helps to understand how concentration polarization varies for reverse osmosis, nanofiltration and ultrafiltration (UF). Indeed, there is a small intersection area for NF and RO for comparable applications such as brackish water softening and removal of micropollutants. UF is clearly differentiated from them with a higher concentration polarization that results from the removal of large organic molecules (such as humic acids and polysaccharides) which may result in fouling due to a cake layer formation adjacent to the membrane. NF will also experience a higher concentration polarization when fouling occurs, mainly when applications involve the presence of low molecular weight humic acids and, moreover, the presence of natural organic matter. This situation is applicable to pilot- and full-scale NF treatment plants that receive pre-treated surface water feeds and secondary effluents of wastewater treatment plants.

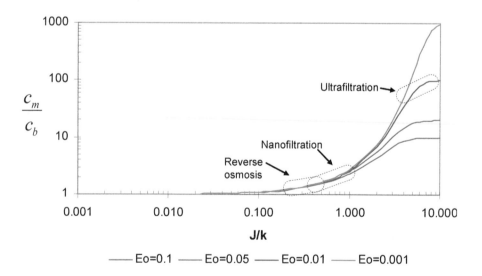

Fig. 3.2: Variation of concentration polarization for RO, NF and UF

3.4.2   Membrane characterization and fouling

The pure water permeability at 20°C for clean and fouled (NF-200 and NF-90) membranes is shown in Table 3.4. The contact angles of clean NF-200 and NF-90 were measured as 37.5° and 58°, respectively (Table 3.4). These measurements indicated that the clean NF-200 membrane was less hydrophobic compared to NF-90. Furthermore, there was apparently no change in membrane hydrophobicity due to alginate fouling. In the case of fouled membranes, significant changes in hydrophobicity were observed in a few studies. For instance, Xu et al. (2005), found a marked increase in hydrophobicity of fouled NF-200 membranes while the hydrophobicity of NF-90 decreased a little. However, they used secondary effluent as a foulant and the DOC delivered to the membrane severely fouled the surfaces, with delivered DOC ranging between 1.5–2.5mg/cm². In this study, the DOC delivered by sodium alginate was approximately 0.14–0.19mg/cm² (15% flux decline). The membrane surface charge, in terms of zeta potential, was analysed at pH 7 for the 10mM KCl solution.

Table 3.4: Comparison of clean and fouled membrane with sodium alginate

Characteristic	NF-200		NF-90	
	Clean	Fouled	Clean	Fouled
Contact angle (degree)	37.5	39.3	58	54.2
Zeta potential (mV)	-10.8	-26.3	-48	-38.2
Pure water permeability PWP (L/m²-day-kPa)	1.01	0.86	2.23	1.90

The clean NF-200 and NF-90 membranes showed zeta potentials of -10.8mV and -48mV, respectively. The zeta potential of fouled NF-200 membranes was observed to increase to -26.3mV while that of the fouled NF-90 membrane decreased to -38.2mV. The foulant layer significantly increased the negative charge of the NF-200 membrane while the more negatively charged NF-90 membrane became slightly less negative after fouling; a possible explanation of this is given subsequently. Salt rejection tests were carried out simulating the standard conditions indicated by the membrane datasheets. The average magnesium sulphate salt rejections by clean and fouled NF-200 membranes were 96.3% and 96.5%, respectively. The salt rejections by clean and fouled NF-90 membranes were 98.4% and 98.7%, respectively.

Fouling was accomplished with a hydrophilic anionic polysaccharide, sodium alginate [(NaC_6H_7O_6)_n], which contained approximately 0.468mg-C/mg. This model foulant was used to establish 9 to 10.5mg/L DOC in the feed solution that resulted in a specific flux decline of ~15% for both membranes within 6 to 8 hours of filtration (pH 7, 10mM KCl, and 20°C). A mass balance assessment of the system enabled the estimation of the DOC delivered to (retained by) the membranes according to Eq. 3.10, which was 0.14mg/cm^2 for NF-200 and 0.19mg/cm^2 for NF-90 membranes assuming a homogeneous distribution of the foulant layer.

The characterization of fouled membranes showed that there was a noteworthy change in the surface charge after fouling. The foulant layer increased the negative charge of the NF-200 membrane while the more negatively charged NF-90 membrane became slightly less negative after fouling. A possible explanation for this phenomenon may be that in the foulant layer of negatively charged sodium alginate, the acid groups are largely dissociated at neutral pH (Van de Ven et al., 2008), covered the surface of the NF-200 membrane ultimately increasing its negative charge. On the other hand, the foulants may have been trapped in the valleys of the rougher NF-90 clean membrane which reduced the surface roughness and may have left the rough crests exposed. The resulting effect may be a slight decrease in the charge of the fouled NF-90 membranes. According to an atomic force microscopy (AFM) image analysis performed by Xu et al. (2006), the NF-200 membrane surface (mean roughness 5.2nm) is smoother than the NF-90 membrane surface (mean roughness 63.9nm).

3.4.3   Rejection (R_1) after membrane saturation

Membrane NF-200

Rejections by clean NF-200 membranes under steady-state conditions (R_1) at two hydrodynamic conditions are shown in Table 3.5. Percent rejections versus molecular weight, molar volume, equivalent width, molecular length, log D and dipole moment were plotted; the graphs are shown in Figs. 3.3 to Fig. 3.5. The figures show results at the two hydrodynamic conditions, rejections at J/k = 0.5 with a recovery of 3% and rejections at J/k = 0.8 with a recovery of 8%. For ionic compounds, a J/k = 0.8 and a recovery of 8% resulted in higher rejections than a J/k = 0.5 and a recovery of 3%. This result may be due to a higher negative charge generated in the first condition due to a higher recovery, thus enhancing electrostatic repulsion of negatively charged compounds. In the case of neutral compounds, a J/k = 0.8 and a recovery of 8% resulted in lower and higher rejections than a J/k = 0.5 and a recovery of 3%, without a clear trend when considering the margins of errors.

Table 3.5: Steady state compound rejection by NF-200 membranes

Name ID	Classification	Steady state (R_1)		
		J/k =0.5, r = 3%	J/k =0.8, r = 8%	
		NF 200	NF 200	NF 200 fouled
ACT	HL-neutral	67.1	94.1	17.7
PHN	HL-neutral	41.3	69.6	21.4
CFN	HL-neutral	50.0	50.0	61.9
MTR	HL-neutral	47.3	34.5	27.6
PHZ	HL-neutral	52.5	61.7	56.4
SFM	HL-ionic	58.9	71.4	48.8
NPN	HB-ionic	75.6	93.9	79.7
IBF	HB-ionic	75.5	93.8	87.5
CBM	HB-neutral	70.0	72.9	73.0
ATZ	HB-neutral	81.3	83.8	88.0
E2	HB-neutral	63.2	60.5	76.5
E1	HB-neutral	76.4	57.3	79.2
NPL	HB-neutral	83.3	83.3	89.7
BPA	HB-neutral	28.5	45.4	51.0

The results indicate that for the NF-200 membrane there is no defined relation between MW and the rejection of neutral compounds (Fig. 3.3a). This can be attributed to the MWCO of NF-200, assumed to be 300Da, which is greater than the molecular weight of hydrophilic and hydrophobic compounds used in this study (151 to 272g/mol), thus not presenting consistent (high) rejections for all compounds. The removal of ACT was rather unexpected for NF-200. ACT as a low molecular weight compound was expected to show lower rejections than other compounds of its group, but that was not the case. Another compound of interest was BPA; that although having a higher MW compared to other compounds of its group showed lower rejection by the membrane; an explanation of this phenomenon is given later.

Molar volume and molecular length showed a linear increase in rejection of hydrophilic neutral compounds, with acetaminophen being an anomaly in the trend. On the other hand, no relation between rejection and molar volume or molecular length was observed for hydrophobic neutral compounds (Fig. 3.3b, 3.4b). Equivalent width did not correlate with rejection of hydrophilic neutral compounds, but there was some relationship evident for hydrophilic and hydrophobic compounds (Fig. 3.4a). No relationships were observed for molecular width and depth (not showed in figures).

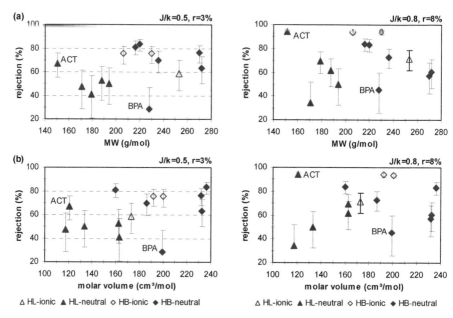

Fig. 3.3: Rejection (R_1, NF-200) vs. compound properties: (a) MW, (b) molar volume
HL = Hydrophilic, HB = Hydrophobic

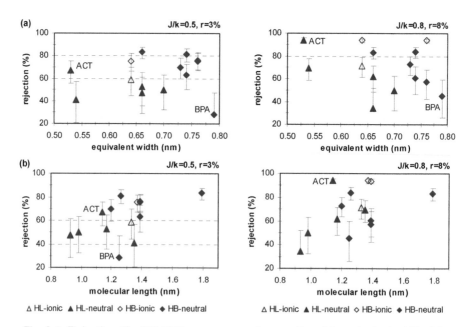

Fig. 3.4: Rejection (R_1, NF-200) vs. compound properties: (a) equivalent width, (b)
molecular length; HL = Hydrophilic, HB = Hydrophobic

Fig. 3.5a shows that log D described rejections of hydrophobic neutral compounds only; in this case, nonylphenol was an exception due to its higher initial concentration. However, BPA owed its low rejection to its high log D, 3.86 (comparable to log K_{ow}, 3.36) with possible partitioning of this compound occurring through the NF-200 membrane due to hydrophobic interactions. It appears that high values of dipole moment had an impact on the rejection of the hydrophilic compounds: metronidazole and sulphamethoxazole (Fig. 3.5b).

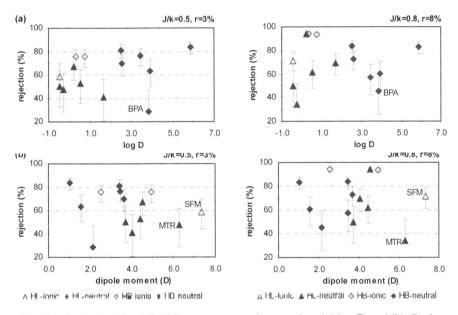

Fig. 3.5: Rejection (R_1, NF-200) vs. compound properties: (a) log D and (b) dipole moment; HL = Hydrophilic, HB = Hydrophobic

The observations regarding MW, size and MWCO suggest that dimensional parameters are important for molecules with different molecular structures. The overall rejection (average steady-state rejection) of NF-200 was ~62% for hydrophobic and hydrophilic neutral compounds; solute permeations across the relatively looser NF-200 membrane may be attributed to facilitated diffusion through the membrane pores, ultimately lowering the overall rejection. The low rejection exhibited for low molecular-weight compounds was observed by other authors using NF-200 or other membranes with similar characteristics. For instance, most of the relatively smaller sized organic contaminants used in a study by Agenson and Urase (2007) were rejected in the range of 23-65% by a clean aromatic polyamide hydrophilic membrane having an MWCO of 350Da. They observed no significant influence of molecular weight or width on rejection, however, they found better correlation with molecular width in the case of larger solutes. Kimura et al. (2004), while testing rejection of neutral EDCs and PhACs, reported 44% rejection of caffeine but only 10% rejection of phenacetine by an SC-3100 membrane (MWCO 200-300 Da). They also demonstrated that there was no significant relationship between the molecular weight of the tested compounds and their corresponding rejections by this membrane. Nevertheless, this study noted that a relatively larger compound, primidone (MW 218g/mol), was well rejected up to 85%, most likely attributed to a sieving phenomenon. In different studies, primidone was also found to be rejected at around 90% by an NF-200 clean membrane (Xu et al., 2005; Xu et al., 2006).

The small group of hydrophobic ionic compounds, comprising naproxen and ibuprofen, exhibited relatively stabilized rejections ranging from approximately 76 to 94% by NF-200 clean membranes (Table 3.5). The hydrophilic ionic compound sulphamethoxazole showed a rejection between 59 to 71% by NF-200 clean membranes (Table 3.5). As a hydrophilic ionic compound, sulphamethoxazole most likely did not adsorb onto the membrane surface, and hence the measured rejection efficiency was similar to that calculated based on the initial feed concentration. The relatively higher degree of rejection of this compound, compared to the neutral compounds, may be attributed to the electrostatic repulsion in addition to the steric hindrance mechanism (Ozaki and Li, 2002; Xu et al., 2006; Bellona et al., 2004; Nghiem et al., 2005). Sulphamethoxazole is negatively charged at pH 7, therefore it could have interacted with the negative surface charge of the NF-200 membrane in addition to being influenced by its molecular size. Naproxen and ibuprofen, two hydrophobic ionic compounds, were rejected at higher degrees, apparently pointing mainly to electrostatic repulsion and less to polarity effects. This is somewhat similar to the findings of Van der Bruggen et al. (1999) who hypothesized that the charge effect is potentially more important for smaller sized ionic molecules than the effect of the membrane pore. Retention of naproxen and ibuprofen by the NF-200 membrane can be inferred as a combined influence of steric hindrance and electrostatic repulsion. Rejection of hydrophobic ionic compounds, among other organic micropollutants, has been probed in several investigations. For instance, Bellona and Drewes (2005) observed approximately 75% rejection of ibuprofen at pH 7 by NF-200 at a recovery of 3%, which coincides with the current findings of this research.

Membrane NF-90

Rejection by clean NF-90 membranes under steady state conditions (R_1) at two hydrodynamic conditions are shown in Table 3.6. Percent rejection versus molecular weight, molar volume, equivalent width, molecular length, log D and dipole moment were plotted; the graphs are shown in Fig. 3.6 to Fig. 3.8. The figures show results at the two hydrodynamic conditions, rejection at J/k = 0.5 with a recovery of 3% and rejection at J/k = 0.8 with a recovery of 8%. For ionic compounds, a J/k = 0.8 and a recovery of 8% resulted in higher rejection (98.5–99%) than a J/k = 0.5 and a recovery of 3% (94.4–96%), this may have resulted from a higher negative charge generated in the first condition due to higher recovery, thus enhancing electrostatic repulsion of negatively charged compounds. In case of neutral compounds, a J/k = 0.8 and a recovery of 8% resulted in lower and higher (62.4–99%) rejections than a J/k = 0.5 and a recovery of 3% (71.2–96%), without a clear trend when considering the margins of errors.

Table 3.6: Steady state compound rejection by NF-90 membrane

Name ID	Classification	Steady state (R_1)		
		J/k =0.5, r = 3%	J/k =0.8, r = 8%	
		NF 90	NF 90	NF 90 fouled
ACT	HL-neutral	71.2	62.4	81.0
PHN	HL-neutral	75.0	70.9	76.0
CFN	HL-neutral	80.8	80.8	91.0
MTR	HL-neutral	82.5	88.3	90.1
PHZ	HL-neutral	85.0	95.6	93.9
SFM	HL-ionic	94.4	98.5	96.5
NPN	HB-ionic	96.0	99.0	96.5
IBF	HB-ionic	96.0	99.0	97.1
CBM	HB-neutral	90.8	98.2	94.5
ATZ	HB-neutral	95.0	97.8	97.1
E2	HB-neutral	90.9	95.3	97.8
E1	HB-neutral	90.3	92.2	96.9
NPL	HB-neutral	90.3	97.8	97.6
BPA	HB-neutral	90.4	95.0	94.6

With regard to NF-90, the rejection of hydrophilic neutral compounds correlated well with molecular weight and equivalent width, as can be seen in Figs. 3.6a and 3.7a. It is to be noted that rejection of hydrophilic neutral compounds increased approximately linearly with molecular weight. However, the rejection of compounds having molecular weight (MW 206–272g/mol) greater than the MWCO of NF-90 (200Da) were not consistent with a linear increase, but showed higher rejections. Furthermore, the evidence that these compounds were rejected by more than 90% supports the

MWCO of the NF-90 membrane. Similar performance was found for an RO membrane (Kimura et al., 2003b and 2004). High rejection of hydrophilic neutral solutes by NF-90 was also reported by others (Xu et al., 2005 and 2006).

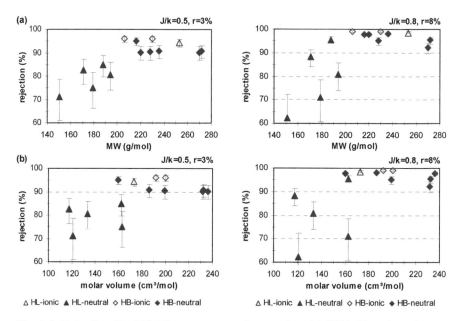

Fig. 3.6: Rejection (R_1, NF-90) vs. compound properties: (a) MW, (b) molar volume
HL = Hydrophilic, HB = Hydrophobic

With regard to the dimensional parameters, the rejection of hydrophilic neutral compounds appeared to be notably influenced by molecular width, depth and equivalent molecular width. Therefore, for the hydrophilic neutral compounds, steric hindrance appeared to be the prevailing rejection mechanism by a clean NF-90 membrane. As opposed to the case of the NF-200 membrane, the hydrophilic ionic compound sulphamethoxazole was rejected by 94% to more than 98% by clean NF-90 membranes. The increased degree of rejection may be ascribed to the negative surface charge as well as to the lower MWCO of NF-90. However, sulphamethoxazole was predominantly retained by NF-90 due to electrostatic repulsion and the sieving effect, consequently the retention was nearly complete. Hence, it can be concluded that pharmaceutical retention by tight NF membranes such as NF-90 appears to be principally governed by steric interaction. The small group of hydrophobic ionic compounds, comprising naproxen and ibuprofen, exhibited relatively stabilized rejections ranging from approximately 96 to 99% by clean NF-90 membranes (Table 3.6). Negative ionic species, naproxen and ibuprofen, were rejected by electrostatic repulsion with the negatively charged membrane, which probably inhibited their adsorption onto the membrane. Moreover, these compounds probably have a low affinity for the membrane polymer, as suggested by their relatively low log D values with no effect on rejection.

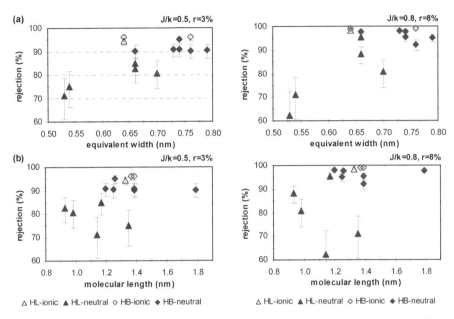

Fig. 3.7: Rejection (R_1, NF-90) vs. compound properties: (a) equivalent width, (b) molecular length; HL = Hydrophilic, HB = Hydrophobic

Dipole moment did not appear to have any influence on rejection of sulphamethoxazole, naproxen and ibuprofen (see Fig. 3.8b). In the case of clean NF-90 membranes, complete to near-complete rejection of naproxen and ibuprofen may be attributed to the coupled effects of steric hindrance and electrostatic repulsion. The NF-90 membrane is even more negatively charged than the NF-200 membrane; hence, the adsorptive interaction of the surface with the ionic compounds can probably be ignored.

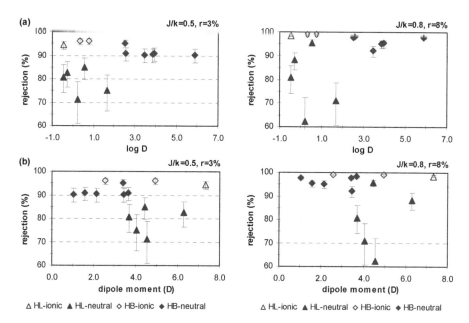

Fig. 3.8: Rejection (R$_1$, NF-90) vs. compound properties: (a) log D and (b) dipole moment; HL = Hydrophilic, HB = Hydrophobic

3.4.4   Rejection (R$_2$) with adsorption

Rejections (R$_2$) of hydrophobic neutral compounds (carbamazepine, atrazine, 17β-estradiol, estrone, nonylphenol and bisphenol A) by the NF-200 membrane varied from 50 to 92%, while that evaluated for NF-90 ranged from 91 to 99% (Table 3.7). In fact, hydrophobic neutral organics can adsorb to the membrane because of moderate (~2) to high (>3.5) log K$_{ow}$ or log D. Fig. 3.9 shows the relationship between hydrophobicity, compound rejection and compound adsorption onto the membranes. There was apparently no correlation between compound rejection and log K$_{ow}$ (Fig. 3.9a, 3.9b); however, a good correlation existed between mass adsorption and log D (Fig. 3.9c, 3.9d). These observations are similar to that of Kiso et al. (2001b) that also found poor correlation between rejection of aromatic pesticides (including atrazine) and log K$_{ow}$, however, their study demonstrated a good correlation between adsorption and log K$_{ow}$. Although adsorption to the membrane was significant, hydrophobic neutral compounds were ultimately rejected due to the size exclusion mechanism, which also explains why these compounds were rejected at higher efficiencies by NF-90 compared to the NF-200 membrane.

Table 3.7: Compound rejection with adsorption by NF-200 and NF-90 membranes

Name ID	Classification	Rejection with adsorption (R_2)					
		J/k =0.5, r = 3%			J/k =0.8, r = 8%		
		NF 200	NF 90	NF 200	NF 200 fouled	NF 90	NF 90 fouled
ACT	HL-neutral	68.5	75.2	95.8	21.4	67.4	83.7
PHN	HL-neutral	50.4	80.0	79.6	45.7	75.9	80.4
CFN	HL-neutral	62.7	84.8	64.6	69.2	85.8	93.3
MTR	HL-neutral	53.7	83.5	43.0	29.0	89.8	91.5
PHZ	HL-neutral	60.4	85.9	72.5	65.7	96.5	95.0
SFM	HL-ionic	61.6	94.5	74.1	52.4	98.5	96.5
NPN	HB-ionic	76.8	96.2	95.1	82.4	99.2	97.1
IBF	HB-ionic	77.3	96.2	95.6	91.4	99.2	97.6
CBM	HB-neutral	78.8	91.3	81.6	80.4	98.6	95.9
ATZ	HB-neutral	88.6	95.7	90.1	94.5	98.4	98.2
E2	HB-neutral	80.6	92.7	77.9	95.1	97.1	98.7
E1	HB-neutral	92.2	93.0	73.1	92.0	94.8	98.0
NPL	HB-neutral	91.7	91.3	91.7	97.6	98.8	98.8
BPA	HB-neutral	50.4	91.5	67.2	80.8	96.6	96.5

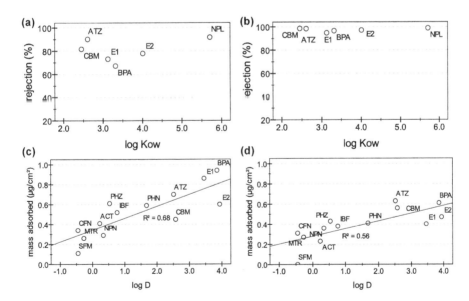

Fig. 3.9: Relationship between rejections (R_2), adsorption and hydrophobicity (J/k=0.8, r=8%): (a) rejection of HB neutral compounds vs. log K_{ow} (NF-200), (b) rejection of HB neutral compounds vs. log K_{ow} (NF-90), (c) Adsorption per unit area vs. log D (NF-200), (d) Adsorption per unit area vs. log D (NF-90)

3.4.5   Rejection, adsorption and hydrogen bonding

It was reported that many hydrophobic trace organics possess some hydrogen-bonding capacity which may be conducive to the adsorptive mechanism, and that hydrogen bonding and hydrophobic interaction can act independently or concurrently. Also, in the latter case, it is often difficult to separate the two independent effects (Nghiem et al., 2002, Schafer et al., 2005). The experimental results, as discussed so far, show that adsorption of hydrophobic neutral compounds occurred in the membrane, but they could not be correlated to hydrogen bonding as shown in Fig. 3.10, and rejection (R_2, in parentheses) did not show any correlation either. A method to indicate a molecule's hydrogen-bonding ability depends on whether the molecule has or does not have atoms acting as donors or acceptors of hydrogen bonds. Another method is the use of molecular electrostatic potentials (Dearden and Ghafourian, 1999; Ghafourian and Dearden, 2004). Adsorption should not be considered as a long-term rejection mechanism because diffusion through the membrane occurs over time. Therefore, as time progresses, rejection is likely to decline after the membrane is saturated after long-term operation. Changes in feed and permeate concentrations, however, were not monitored as a function of time in the current research.

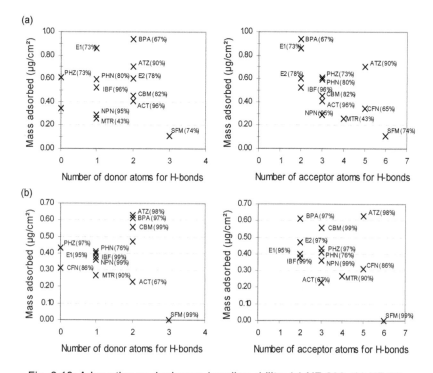

Fig. 3.10: Adsorption vs. hydrogen-bonding ability: (a) NF-200, (b) NF-90

3.4.6   Rejection of compounds by fouled membranes

The steady-state rejections (R_1) and rejection with adsorption (R_2) of selected PhACs and EDCs by fouled NF-200 and NF-90 membranes, as compared to the clean ones, were presented in Tables 3.5, 3.6 and 3.7. A decreasing trend was observed in R_1 rejections of hydrophilic neutral compounds when the NF-200 membrane was fouled with sodium alginate. For instance, the rejection of phenacetine, metronidazole and phenazone by fouled NF-200 membranes decreased by 69%, 20% and 9%, respectively, as compared to that by clean membranes. Moreover, as already mentioned, the rejection of acetaminophen was an anomaly in rejection trends of hydrophilic neutral compounds by the clean membrane and, hence, the decrease in rejection by the fouled membrane could be considered as less prominent (9–20%). On the other hand, the largest MW compound of the group, caffeine, exhibited an increased rejection. Decrease in the rejection of hydrophilic neutral compounds by NF-200 may be explained by the phenomenon of "cake-enhanced concentration polarization" (Hoek et al., 2002 and 2003; Lee et al., 2006). It is hypothesized that the foulant layer (cake) hinders the back diffusion of solutes from the membrane surface to the bulk solution. The accrued solutes at the membrane surface enhance the concentration gradient across the membrane, leading to an increase in permeate concentration and thereby a decrease in the observed rejection. This is true only for small compounds (phenacetine, phenazone, metronidazole) rejected by NF-200. The ionic compound sulphamethoxazole presented the same rejection for both the fouled and clean NF-90 membranes. Membrane fouling increased the negative charge of the NF-200 membrane and therefore was expected to exhibit an increased rejection of the negatively charged compound. However, the alginate macromolecules have a stretched and less compact configuration above pH 5 due to increased electrostatic repulsion between carboxyl groups and thus form a comparatively sparser fouling layer (Lee et al., 2006). The gradual build-up of the retained molecules of sulphamethoxazole, facilitated by convective transport, resulted in an elevated concentration gradient at the membrane surface. Solute transport away from the membrane surface by back diffusion was hindered by the foulant layer which ultimately favours solute transport across the membrane.

The rejections with adsorption (R_2) of hydrophobic ionic compounds, naproxen and ibuprofen, by fouled NF-200 membranes, decreased by 13% and 4%, respectively, when compared to the clean membranes. The rejection values (R_1) for fouled membranes under steady-state conditions were close to rejections with adsorption (R_2). Similar to the rejection mechanism of sulphamethoxazole, the decreasing trend of rejection of naproxen and ibuprofen, although relatively less pronounced, might be attributed to the hindered back diffusion and subsequent solute transport through the membrane. On the other hand, rejection of the hydrophobic neutral compounds atrazine, 17β-estradiol, estrone, nonylphenol and BPA were observed to be increased by 5, 22, 26, 6 and 20% respectively, by the fouled membrane compared to the clean NF-200.

Although the hydrophobicity of the fouled NF-200 membrane was not notably changed, adsorption of the hydrophobic neutral compounds was found to increase with respect to the clean membrane, most likely because of the additional foulant layers (Fig. 3.11). Furthermore, unlike the ionic compounds, the hydrophobic neutrals might have clustered together preferentially in the interior of macromolecules of the foulant layer, impeding their interaction with the membrane surface. Thus, the enhanced rejection of hydrophobic neutral compounds by the fouled NF-200 membrane may be attributed to the incipient interaction of the solutes with the membrane resulting in less partitioning and diffusion across the membrane. A similar observation was drawn by Xu et al. (2006) who observed increased rejection of chloroform, bromoform and trichloroethylene by polyamide NF and ULPRO membranes fouled with micro-filtered secondary effluent when compared to the clean membranes.

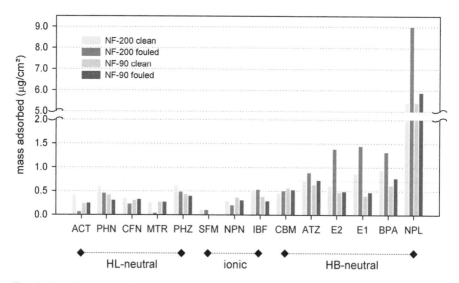

Fig. 3.11: Adsorption of compounds to clean and fouled membranes (J/k=0.8, r=8%)

With regard to the NF-90 membrane, there was no clear alteration of membrane hydrophobicity due to fouling. The rejection of hydrophobic compounds by the clean membrane was already nearly complete and was not observed to be distinctly affected by fouling. In this case, however, hydrophilic neutral compound rejections increased up to 30% due to fouling.

Although the rejection was not monitored over time, an initial decline might have occurred; nevertheless, the improved rejection was a phenomenon due to the intrinsic behaviour of the "tighter" NF-90 membrane. In comparison with clean membranes, fouled-membrane rejection values correlated well with the molecular size parameters as can be seen in Fig. 3.12. Thus, the increased rejections indicate that an enhanced sieving effect may dominate rejection mechanisms of hydrophilic neutral compounds by fouled NF-90 membranes.

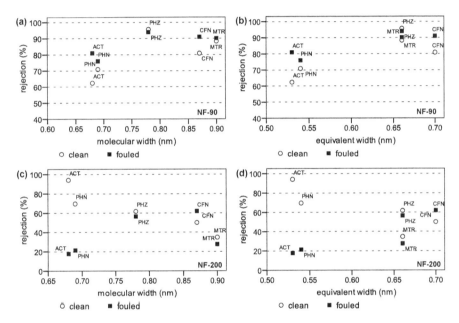

Fig. 3.12: Relationship between rejection (R_1) and size of HL neutral compounds by clean and fouled membranes (J/k=0.8, r=8%): (a) Rejection vs. molecular width (NF-90), (b) Rejection vs. equivalent width (NF-90), (c) Rejection vs. molecular width (NF-200), (d) Rejection vs. equivalent width (NF-200)

3.5   Conclusions

o   Experiments with flat-sheet NF membranes showed comparable values of concentration polarization modulus for organic solutes compared to concentration polarization of NF membrane elements; suggesting that results of flat-sheet membranes may be compared to membrane elements.

o   Rejection of ionic compounds by NF membranes may be mainly attributed to electrostatic repulsion. Electrostatic repulsion between the negative charge of the ionic specie of the solute and the negative charge of the membrane surface was the main mechanism of rejection for ionic compounds.

o   It appears that rejection of hydrophilic neutral compounds increased with increasing molar volume and molecular length, thus steric hindrance may be an important mechanism influencing rejection of this group of compounds.

o   Significant adsorption of hydrophobic neutral compounds to the membranes was observed; adsorption increased almost linearly in relation to log D. Adsorption of ionic and hydrophilic neutral compounds was less significant.

o   A combined effect of size and log K_{ow} (log D) primarily resulted in higher rejection of neutral compounds by clean NF-200 membranes, and nearly complete rejection by clean NF-90 membranes.

o   Alginate fouling of the NF-200 membrane slightly decreased rejection of hydrophilic neutrals as well as hydrophilic and hydrophobic ionic compounds possibly due to restricted back diffusion to the bulk solution and subsequent transport across the membrane.

o   The rejection of hydrophobic neutral compounds by NF-90 was not observed to be distinctly affected by alginate fouling, however, hydrophilic neutral compounds showed increased rejection, and may be attributed to the domination of an enhanced sieving effect.

Chapter 4

Effects of NOM and surrogate foulants on the removal of emerging organic contaminants (PhACs, PCPs, EDCs) with NF membranes

Based on parts of:
- Rejection of emerging contaminants by nanofiltration membranes, High Quality Drinking Water Conference, 9-10 June, 2009, Delft, the Netherlands.
- Is Nanofiltration a Robust Barrier for Organic Micropollutants, 5th IWA Specialized Membrane Technology Conference for Water and Wastewater Treatment, IWA Membrane Technology Conference & Exhibition, 1-3 September, 2009, Beijing, China.

4.1   Introduction

In Chapter 3, the experimental work corresponding to the fouling part was carried out using only sodium alginate as a foulant. In this chapter, the experimental work regarding fouling is more extensive. New organic compounds were included and, more importantly, surrogate and real foulants were used. The filtration experiments considered in this chapter are shown in Table 4.1; the percentages shown next to the names (in parentheses) indicate flux decline degrees.

Table 4.1: Filtration experiments

Name (flux decline)	Membrane	Foulant	Feed water
Clean NF-200	NF-200	None	Synthetic water
Alginate (22%)	NF-200	Sodium alginate	Synthetic water
Dextran (12%)	NF-200	Dextran	Synthetic water
NOM (15%)	NF-200	NOM	Synthetic water
NOM (22%)	NF-200	NOM	Synthetic water
NOM (50%)	NF-200	NOM	Synthetic water
NOM (35%)*	NF-200	NOM	Pre-filtered surface water
Clean NF-90	NF-90	None	Synthetic water
NOM NF-90 (28%)	NF-90	NOM	Synthetic water

* Direct fouling/filtration experiment

4.2   Experimental

4.2.1   Experimental setup and organic contaminants

The experimental setup referred to in this chapter was described in Section 3.3.2. The list of organic contaminants with their respective physicochemical properties is presented in Table 4.2. A compound of particular interest is 1,4-dioxane, a small MW compound that was only used for the filtration tests with NF-90 membranes. In order to understand the retention mechanism, compounds which have a diverse range of properties were chosen in terms of molecular weight, molecular size (length, effective diameter, equivalent width), dipole moment, hydrophobicity (log K_{ow}, log D), and acid dissociation constant (pK_a). The feed solution was prepared from a stock solution with a concentration of 1mg/L per compound. The intended concentration for each compound in the feed solution was 10μg/L.

Table 4.2: Compounds and physicochemical properties

Name	Name ID	MW (g/mol)	log K_{ow}[a]	logD[b] (pH 7)	Dipole moment (debye)[c]	Molar volume[d] (cm^3/mol)	Molec. length (nm)[d]	Equiv. width (nm)[d]	Effe. diam. (nm)[d]	pK$_a$[b]	Class.[e]
Acetaminophen	ACT	151	0.46	0.23	4.55	120.90	1.14	0.53	0.79	10.2	HL-n
Phenacetine	PHN	179	1.58	1.68	4.05	163.00	1.35	0.54	0.89	n.a.	HL-n
Caffeine	CFN	194	-0.07	-0.45	3.71	133.30	0.98	0.70	0.77	n.a.	HL-n
Metronidazole	MTR	171	-0.02	-0.27	6.30	117.80	0.93	0.66	0.75	n.a.	HL-n
Phenazone	PHZ	188	0.38	0.54	4.44	162.70	1.17	0.66	0.83	n.a.	HL-n
1,4-dioxane	DIX	88	-0.27	-0.17	0.00	89.10	0.71	0.59	0.57	n.a.	HL-n
Sulphamethoxazole	SFM	253	0.89	-0.45	7.34	173.10	1.33	0.64	0.89	5.7	HL-i
Fenoprofen	FNP	242	-0.02	0.38	1.88	180.90	1.16	0.83	0.88	4.3	HL-i
Ketoprofen	KTP	254	-0.52	-0.13	3.42	187.90	1.16	0.83	0.87	4.3	HL-i
Naproxen	NPN	230	3.18	0.34	2.55	192.20	1.37	0.76	0.93	4.3	HB-i
Ibuprofen	IBF	206	3.97	0.77	4.95	200.30	1.39	0.64	0.93	4.3	HB-i
Gemfibrozil	GFB	250	4.77	2.30	0.95	221.90	1.58	0.78	1.09	4.9	HB-i
Carbamazepine	CBM	236	2.45	2.58	3.66	186.50	1.20	0.73	0.89	n.a.	HB-n
17β-estradiol	E2	272	4.01	3.94	1.56	232.60	1.39	0.74	0.97	10.3	HB-n
Estrone	E1	270	3.13	3.46	3.45	232.10	1.39	0.76	0.97	10.3	HB-n
Bisphenol A	BPA	228	3.32	3.86	2.13	199.50	1.25	0.79	0.89	10.3	HB-n
17α-ethynylestradiol	EE2	296	3.67	3.98	1.27	225.60	1.48	0.85	1.02	10.3	HB-n

[a] Experimental database: SRC PhysProp Database
[b] ADME/Tox Web Software
[c] Chem3D Ultra 7.0
[d] Molecular Modeling Pro
[e] HL = Hydrophilic, HB = Hydrophobic, n = neutral, i = ionic hydrophobic if log K$_{ow}$ > 2

4.2.2   Foulants

Membranes were fouled with 1) sodium alginate [(C$_6$H$_7$O$_6$Na)$_n$], (Sigma-Aldrich, Germany), a hydrophilic anionic polysaccharide produced by algae and bacteria; and, 2) dextran from *Leuconostoc mesenteroides* [C$_6$H$_{10}$O$_5$]$_n$, (Sigma-Aldrich, Germany); a neutral compound containing complex branched glucans (polysaccharide made of many glucose molecules). Both were prepared as stock solutions with high concentrations. Before each fouling test, the solution was added to the feed tank. The concentration of sodium alginate was 900 mg-C/L and dextran was 300 mg-C/L in the feed solution. A summary of information related to the foulants is presented in Table 4.3. The feed solution with NOM was created by filtering Delft canal water, taken from the canal in front of UNESCO-IHE, and pre-filtered with a 1.2μm pore size capsule filter (Sartopure PP2, Sartorius Gmbh, Germany).

Table 4.3: Compounds and physicochemical properties

Name	Content type	Size range	Source
Sodium alginate	Polysaccharides	12–80kDa	Nghiem et al., 2008a
Dextran	Polysaccharides	9–11kDa	Sigma (manufacturer)
NOM	Polysaccharides Humic/Fulvic acids	300Da–50kDa	LC-OCD analysis

The water quality parameters of the feed solution (pre-filtered Delft canal water) containing NOM are shown in Table 4.4. During the NOM pre-fouling experiments, the feed tank contained this pre-filtered solution. Additionally, a characterisation of NOM by means of liquid chromatography organic carbon detection (LC-OCD) is presented in Fig. 4.1. Humics, building blocks and polysaccharides were the main constituents of the pre-filtered canal water.

Table 4.4: Characteristics of pre-filtered 1.2µm surface canal water

Parameter	Unit	Value
Turbidity	NTU	1.1 - 1.6
DOC	mg/L	19.3 - 19.5
UV_{254}	1/cm	0.58 - 0.74
SUVA	L/mg/cm	0.030 - 0.038
Conductivity	µS/cm	921 - 1134
pH	-	7.9
Fe	mg/L	0.22 - 0.72
Mg^{2+}	mg/L	17.4 - 24.0
Mn	mg/L	0.11 - 0.37
Ca^{2+}	mg/L	119 - 133
Si	mg/L	2.3 - 6.1

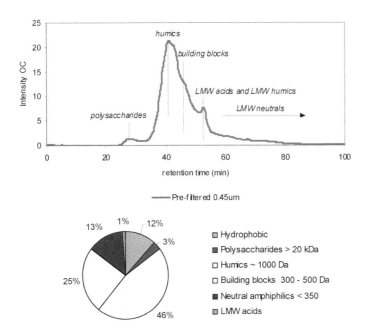

Fig. 4.1: LC-OCD chromatogram and composition of pre-filtered surface water

4.2.3   Compaction, pre-fouling and filtration tests

The general protocol of the filtration experiment is shown in Fig. 4.2. At the beginning of the tests, compaction with clean water was carried out to stabilise membrane permeability. The compaction process was performed for 3 hours. For the pre-fouling and filtration experiments, permeate flow and concentrate flow were set to the hydrodynamic conditions shown in Table 4.5.

After compaction, the clean water solution in the feed tank was changed to a solution containing one of the foulants and 10mM KCl. As mentioned, three different foulants (sodium alginate, dextran and NOM), were utilised for fouling NF-200 membranes. NF-90 membranes were fouled only with NOM. The concentrate was recycled into the feed tank during the pre-fouling test. During the pre-fouling, the concentrate flow was set sufficiently high, approximately twice the concentrate flow of a filtration rejection experiment, in order to guarantee that the fouling cake layer remained almost intact during the filtration rejection tests using the organic compounds. This was verified controlling that the permeate flow was constant during an experiment with clean water at the same hydrodynamic conditions of a rejection test. The permeate flow was constantly controlled to determine the flux decline.

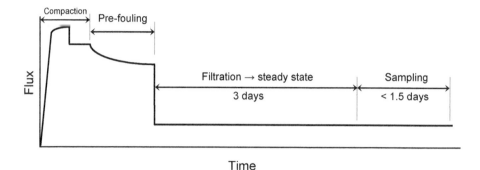

Fig. 4.2: General protocol of the compaction, pre-fouling and filtration experiment

Table 4.5: Hydrodynamic conditions of the experiments

Parameter	Filtration (compound rejection)		Pre-fouling	
Recovery (%)	8		18	
Flux, J (L/m²-h)	13		32	
	NF200	NF90	NF200	NF90
Pressure, P (kPa)	483	345	688	552
U, average cross-flow velocity (cm/s)	5		10	
Mean diffusion coeff., D (cm/s) for organic solutes	6.15E-06		-	
Back diffusion mass transfer coeff., k, for org. sol.	6.7E-04		-	
J/k	0.5		-	

After the pre-fouling test, the fouling solution in the feed tank was replaced by 50L of synthetic water solution at 10mM KCl. The synthetic water solution was adjusted to maintain a pH of 7. A cocktail of compounds prepared from a stock solution was added to the feed solution (synthetic water solution) in order to have an approximate concentration of 10μg/L for each compound. Steady-state conditions of saturation of the membranes were achieved after three days of filtration in a recycle mode with the concentrate and permeate recirculated to the feed tank. After membrane saturation, 4L of permeate were collected over about 1.5 days. Throughout the tests, a solution temperature of 20 ±0.5°C was kept constant.

4.2.4   Analytical methods

As mentioned in Chapter 3, analyses of the compounds were performed at Technologiezentrum Wasser TZW (Karlsruhe, Germany). The new compounds (fenoprofen, gemfibrozil and ketoprofen) were analysed according to the method described in Section 3.3.3 for Group A; and the new compounds (1,4-dioxane and 17α-ethynylestradiol) were analysed according to the method described in that section for Group C.

The pH, conductivity and DOC were measured by the equipment described in Section 3.3.3. The contact angle was measured by a CAM200 optical contact angle and surface tension meter (KSV Instruments, Finland). Additionally, for membrane analyses, attenuated total reflection using an infrared Fourier transform (ATR-FTIR) spectroscopy (FT-IR Spectrophotometer Perkin Elmer 100) was used to identify the functional groups of foulants deposited on the membrane. By using this method, foulants such as humics, protein- and polysaccharide-like substances were differentiated. The surface charge, in terms of the zeta potential, of clean and fouled membranes was quantified using an electrokinetic (streaming potential) analyser (Anton Paar, Austria). The zeta potential analyses were performed at a standard pH of 7 and ionic strength of 10mM KCl.

Salt rejection tests were carried out as separate tests for clean and fouled membranes. The method specified by the manufacturers was adopted to evaluate salt rejection by membranes. A water solution containing 2000mg-MgSO$_4$/L (around 2,400μS/cm) at pH 8 and temperature 25°C was used. The filtration process to evaluate salt rejection was performed for one hour at a recovery of 15% and a pressure of 483 kPa, following the membrane manufacturer's protocol. After the one-hour filtration, the salt rejection was evaluated by measuring the conductivity in the feed solution and in the permeate solution.

4.3   Results and discussion

4.3.1   Membrane fouling

Flux decline after membrane fouling is presented in Fig. 4.3. The flux for
NF-200 fouled by sodium alginate dropped by around 10% within 15
minutes of the start of filtration; after that, the flux gradually and constantly
decreased (Fig. 4.3a).

The flux for NF-200 fouled by dextran presented a slow decline compared to
sodium alginate (Fig. 4.3b). For the NF-200 membrane fouled by NOM, the
flux declined with a trend similar to that of sodium alginate. The flux
dropped by over 10% within 15 minutes and then suddenly increased before
a gradual decrease resumed (Fig. 4.3c). This phenomenon was also observed
when NF-200 was fouled by NOM up to a 50% flux decline (Fig. 4.3d). This
unstable flux decline using the NF-200 membrane was observed in previous
studies as well (Her et al., 2000; Xu et al., 2006). The flux decline of the NF-
90 membrane fouled by NOM is shown in Fig. 4.3e. The flux rapidly
decreased by 22% within a short period of time 2.5 hours. The difference in
the flux decline rate between the two membranes was probably caused by
the difference in roughness of the active layer of the NF-200 and NF-90
membranes, and by the differences in water permeability. The NF-90
membrane has greater roughness, 63.9–76.8 nm, compared to the NF-200
membrane, 5.2 nm (Nghiem et al., 2008b; Xu et al., 2006).

The water quality of the feed and permeates after the NOM fouling
experiments is shown in Table 4.6. The removal of DOC by both the NF-200
and the NF-90 membranes was high, as expected. The remarkable difference
between the NF-200 and NF-90 membranes was the conductivity of the
permeates, 527–630 µS/cm and 163µS/cm for NF-200 and NF-90,
respectively. This occurred because the NF-90 clean membrane has higher
(divalent) ion rejection (~98%) than the NF-200 (~97%).

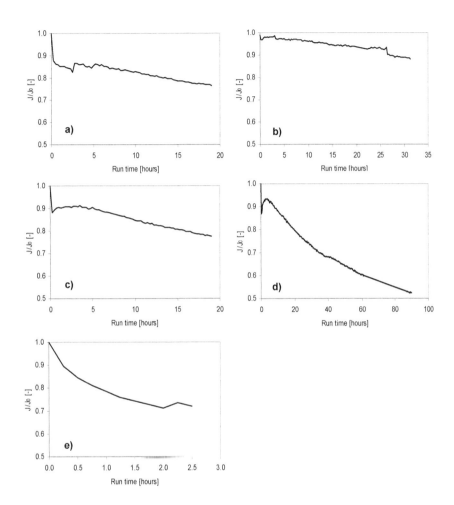

Fig. 4.3: Flux decline, (a) sodium alginate, NF200, (b) dextran, NF200, (c) NOM 22% flux decline, NF200, (d) NOM 50% flux decline, NF200, (e) NOM 28% flux decline, NF90

Table 4.6: Characteristics of feed and permeate

Sample & parameter	NF-200 NOM 22%	NF-200 NOM 50%	NF-90 NOM 28%
Feed (pre-filtered) water			
DOC (mg/L)	19.5	19.4	19.4
UV_{254} (1/cm)	0.71	0.74	0.74
SUVA (L/mg/cm g)	0.036	0.038	0.038
Conductivity (μS/cm)	921	1041	1041
pH	-	7.9	7.9
Ca (mg/L)	119	126	126
Fe (mg/L)	0.22	0.29	0.29
Mn (mg/L)	0.37	0.29	0.29
Mg (mg/L)	17.4	20.3	20.3
Si (mg/L)	6.1	5.6	5.6
Permeate water			
DOC (mg/L)	0.27	0.29	0.28
UV_{254} (1/cm)	0.019	0.006	0.005
SUVA (L/mg/cm)	0.070	0.021	0.018
Conductivity (μS/cm)	527	630	163
Ca (mg/L)	32.6	-	0.93
Fe (mg/L)	0.039	0.03	0.043
Mn (mg/L)	0.16	0.099	0.056
Mg (mg/L)	3.2	4.31	0.26
Si (mg/L)	4.2	1.34	2.06

The surface charge measured as the zeta potential of clean and fouled NF-200 and NF-90 membranes is shown in Table 4.7.

Table 4.7: Zeta potential measurements of clean and fouled membranes

Membrane	Zeta potential (mV)
NF-200	
Clean	-28 (±3)
Alginate	-24 (±2)
Dextran	-14 (±4)
NOM (22% flux decline)	-30 (±4)
NF-90	
Clean	-32 (±3)
NOM (28% flux decline)	-22 (±2)

The contact angles of clean and fouled membranes are shown in Fig. 4.4. It was observed that foulants increased the hydrophobicity of the membrane. The deposit of foulants on the membrane, once dried at room temperature, created a compact layer that decreased permeation of the drop of water that was placed when the contact angle was measured. A membrane does not necessarily become more hydrophilic when a "hydrophilic" foulant covers the membrane, at least not in dry conditions used to measure the contact angle with the sessile drop technique. Therefore, contact angle measurements for one specific type of membrane are more an indication of lower water permeability (hydrophobicity) than the affinity of water to the original foulant; hence, although alginate and dextran are catalogued as hydrophilic substances, the results clearly show an increase in hydrophobicity of the membranes.

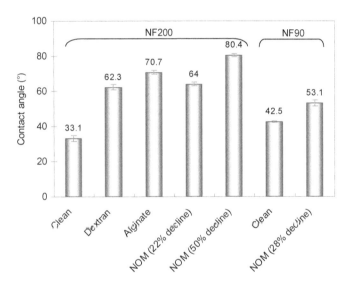

Fig. 4.4: Contact angle of clean and fouled membranes

The results of salt rejection with $MgSO_4$ are shown in Fig. 4.5. Salt rejections of clean and fouled NF-200 membranes presented slight differences, except for the NOM fouled membranes. NF-90 membranes also presented slight differences in the salt rejection between clean and fouled membranes. Although there were small differences, it is important to mention that, in general, the NOM fouling layer slightly increased salt rejection.

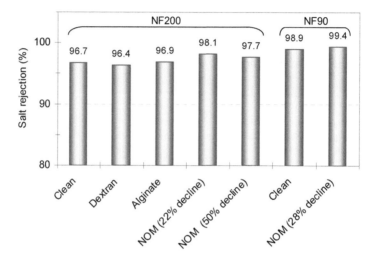

Fig. 4.5: Salt rejection ($MgSO_4$) of clean and fouled membranes

Table 4.8 shows the organic groups that can be identified by ATR-FTIR spectra analysis. According to the references listed in the table, the wave number of around 1720 cm^{-1} represents carboxylic groups (C=O) and reflects humic acid fouling (Jarusutthirak et al., 2002; Xu et al., 2006). The peaks of around 1640 and 1540 cm^{-1} are attributed to amide I band (-NH$_2$) and amide II band (-NH) and can be associated with protein fouling (Howe et al., 2002; Kim and Dempsey, 2008; Xu et al., 2006). A wave number of around 1034 cm^{-1} represents C-O and is associated with polysaccharide fouling under the conditions that there are broad peaks at 2900 cm^{-1} and from 3000 to 3400 cm^{-1} (Howe et al., 2002; Xu et al., 2006).

Table 4.8: Identification of organic groups with ATR-FTIR spectra

Wave numbers (cm^{-1})	Groups	(Howe et al., 2002)	(Xu et al., 2006)	(Kim and Dempsey, 2008)	(Jarusutthirak et al., 2002)
1720	Humic acids	(1725cm^{-1}) C=O (Carboxylic acids)	(1740cm^{-1}) Humic acids		(1740cm^{-1}) Carboxylic groups Humic / fulvic acids
1635 - 1640	Amides	(1653cm^{-1}) Amide I band	(1650cm^{-1}) -NH$_2$	(1640cm^{-1}) C=O stretching in amide groups.	(1640cm^{-1}) -NH$_2$
1535 - 1540		(1543cm^{-1}) Amide II band	(1550cm^{-1}) -NH	(1647cm^{-1}) N-H bonding vibration peak tailing.	(1540cm^{-1}) -NH
1035 - 1065	Polysac-charides	(1034cm^{-1}) C-O / Si-O Polysac. contain a significant number of hydroxyl groups, which exhibit a broad rounded absorption band above 3000cm^{-1} (hydroxyl).	(1030-1040cm^{-1}) -SO, -CO, or -SiO Organic sulfonic acids..P lysac. contain a number of -CH and -OH groups which exhibit a peak at 2930 cm^{-1}, and broad absorption bands at 3000 and 3400 cm^{-1}.	(1010, 1070cm^{-1}) C-O-C, C=O Polysac. or polysac-like substances.	(1040cm^{-1}) Polysac. 1170, 1125, and 1040cm^{-1} are derived from anionic surfactant degradation products.
770 - 825		-		(800cm^{-1}) Glycosidic linkage in polysac.	

Figures 4.6, 4.7 and 4.8 show the plotted differential spectra between the clean membranes and the fouled membranes. The NOM (22% flux decline) and NOM (50% flux decline) on the NF-200 membranes presented in both cases strong peaks at 1035 cm^{-1}, which represent fouling due to polysaccharide-like substances. Also, amide groups were identified at 1640 cm^{-1} and 1535 cm^{-1}. The foulants accumulated more after 50% flux decline, as can be seen in the differential spectra of NOM (Fig. 4.6). However, humic

acids (observed at 1720 cm^{-1}) did not show strong peaks, suggesting that humic acids were not the main foulants of the NF-200 membranes.

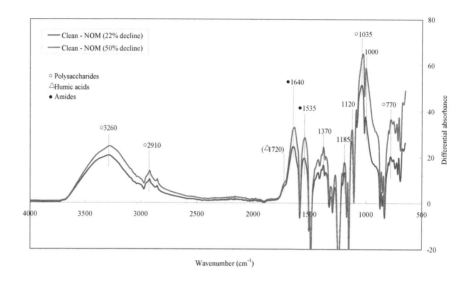

Fig. 4.6: Differential ATR-FTIR spectra for NF-200 membrane fouled with NOM

Figure 4.7 shows ATR-FTIR spectra for the NF-200 membrane fouled with alginate and dextran. The degree of differential absorbance in both cases was lower than that of NOM. As the main component of alginate and dextran is polysaccharides, the peak at 1045 cm^{-1} confirmed the existence of polysaccharides on the fouled membranes. Although the compositions of alginate $(C_6H_7O_6Na)_n$ and dextran $(C_6H_{10}O_5)$ do not contain amide groups in their chemical structures, amides were identified at 1650 cm^{-1} and 1540 cm^{-1}. These observations occurred due to the natural origin of both compounds, sodium alginate and dextran are processed from algae and bacteria, respectively; therefore, parts of protein-like organic matter are present in their structures. Regarding the NF-90 membrane, the peaks on the NOM fouled NF-90 membrane were similar to those identified for the NF-200 membrane fouled by NOM. This indicates that polysaccharides were also the main foulant on the NF-90 membrane.

Although the hydrophobicity of membranes was increased by fouling (contact angle results), polysaccharides are "hydrophilic"; hydrophobicity increased apparently due to compact layer of fouling that increases contact angles (hydrophobicity).

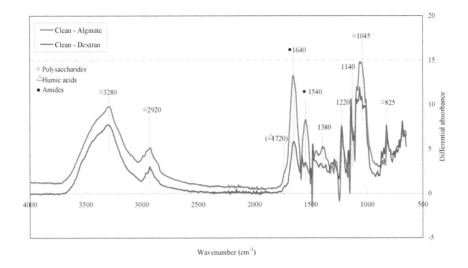

Fig. 4.7: Differential ATR-FTIR spectra for NF-200 membrane fouled with dextran and alginate

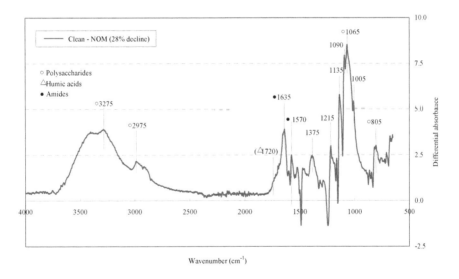

Fig. 4.8: Differential ATR-FTIR spectra for NF-90 membrane fouled with NOM

4.3.2   Rejection of neutral compounds by NF-200

All rejections were calculated using Eq. 3.4 after membrane saturation (steady state). Rejections after experiments using clean and fouled membranes are shown in Fig. 4.9 to 4.15. Compounds have been sorted in ascending order based on molecular weight (Fig. 4.9), molecular length (Fig. 4.10), equivalent width (Fig. 4.11), effective diameter (Fig. 4.12), log K_{ow} (Fig. 4.13) and dipole moment (Fig. 4.14) in order to identify visible correlations between removal and individual physicochemical properties.

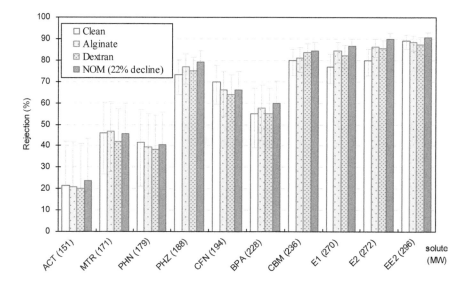

Fig. 4.9: Neutral compound rejections (MW) by NF-200 membranes

NF-200 membranes fouled with sodium alginate showed higher rejection (up to 9%) but also lower rejection (6%) for two cases (caffeine, phenacetine). Membranes fouled with dextran showed a mixed tendency to increase or decrease rejection with changes ranging from -9% to 6%. On the other hand, membranes fouled with surface water (Delft canal) NOM mainly increased rejection up to 11%, except for caffeine which showed reduced (-6%) rejection. Nevertheless, the mentioned variations are not significant in terms of the margin of error determined by the laboratory during the analysis of compounds (~15% in concentrations). Therefore, after observing the rejection of clean and fouled NF-200 membranes, the impact of fouling (for alginate and dextran) was found not to be significant in decreasing or increasing rejection. The impact of NOM fouling did have a slightly significant effect in rejection (decreasing or increasing). It is hypothetized that rejection increased due to an effective foulant layer created by humics, polysaccharides and divalent ions (Ca^{2+}, Mg^{2+}) in the feed water while pre-fouling the membrane.

Regarding the relation of MW and rejection, a compound with high MW is expected to be better removed than a low MW compound. However, that was not always the trend; for instance, phenacetine (PHN) showed lower rejections than metronidazole (MTR). Bisphenol A (BPA) and caffeine (CFN) showed lower rejections than phenazone (PHZ), see Fig. 4.9. Therefore, it is not always true that a high MW compound will be better rejected than a low MW compound.

A more interesting discussion arises from the observations of Figures 4.10 to 4.14 related to other physicochemical properties of compounds.

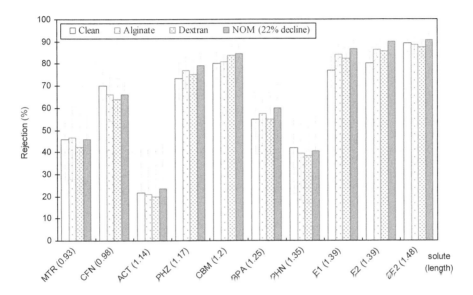

Fig. 4.10: Neutral compound rejections (length) by NF-200 membranes

To understand the differences in rejection between compounds of similar MW, it is necessary to consider the geometric configuration of compounds. Fig 4.10 and 4.11 can help with this; lower rejections of CFN were probably due to its smaller molecular length compared to PHZ. However, lower rejection of PHN compared to MTR cannot be explained by differences in molecular length. The same is true for rejection of acetaminophen (ACT) compared to CFN. A second geometrical size descriptor is needed to understand rejection; that descriptor is equivalent width. Fig 4.11 shows that rejection is proportional to the increase in equivalent width, except for two cases (i.e. CFN and BPA). However, CFN is influenced by its low molecular length compared to other compounds (Fig. 4.10). Another size descriptor considered to explain rejection was the effective diameter (defined in Section 2.3.1). Fig. 4.12 shows that rejection is proportional to the effective diameter, except for ACT, BPA and PHN. Differences in rejection for those compounds cannot be related only to the effect of one size descriptor.

Nonetheless, BPA is a special case that needs more explanation than 'geometric size' to support its moderate rejection.

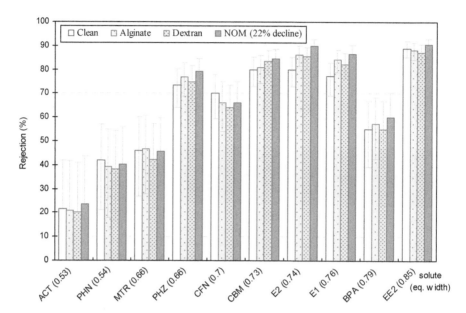

Fig. 4.11: Neutral compound rejections (eq. width) by NF-200 membranes

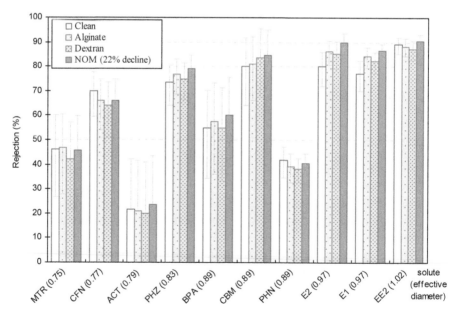

Fig. 4.12: Neutral compound rejections (effective diameter) by NF-200 membranes

An explanation for the moderate rejection of BPA can be its hydrophobicity (log K_{ow} = 3.32, Fig. 4.13). The size (length and equivalent width) and log K_{ow} facilitate a partitioning movement of BPA through the membrane. Partitioning, however, has not been found to be an influential rejection/transport mechanism for more hydrophobic compounds such as estrone (E1), 17β-estradiol (E2) and 17α-ethynylestradiol (EE2) due to their greater size mainly in terms of molecular length and effective diameter (Figs. 4.10 and 4.12) and contributions of equivalent width (Fig. 4.11). Therefore, for the particular case of NF-200 membranes, BPA can be catalogued as a special case that combines size and hydrophobic conditions that may favour its partitioning (diffusion) through the membrane.

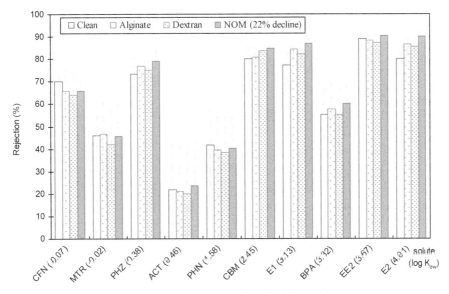

Fig. 4.13: Neutral compound rejections (log K_{ow}) by NF-200 membranes

Finally, Fig. 4.14 presents dipole moments arranged in ascendant order. A first look at the figure may possibly lead to an interpretation that high dipole moments are associated with low rejections; for example, ACT and MTR, compounds with dipole moments 4.55 and 6.3, respectively, showed lower rejections than CFN and PHZ. However, these observations can be misleading because ACT and MTR are smaller compounds than CFN and PHZ. Therefore, it is important to consider appropriate physicochemical variables to explain rejections. This task can be exhausting when many variables and more compounds are involved and, thus, it is necessary to define which variables are more important than others in order to determine *a priori* an estimated rejection of a new compound.

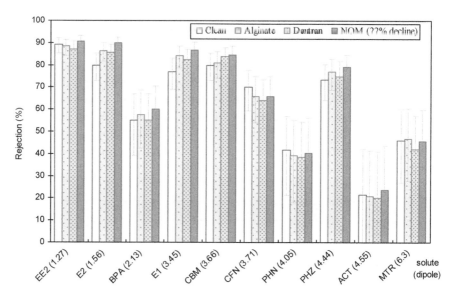

Fig. 4.14: Neutral compound rejections (dipole moment) by NF-200 membranes

It was mentioned that NOM from Delft canal water tended to increase rejection of neutral compounds. The question was whether the degree of fouling was capable of influencing the magnitude of rejection. To explicate, different fouling intensities identified by the extent of flux decline were evaluated; the results are shown in Fig. 4.15. It was found that fouling that produced 50% flux decline increased rejections up to 14%, however some exceptions were still present; CFN presented no change compared to the clean membrane, and rejection of PHN decreased by 4%. Although we have been able to identify changes in rejection due to the intensity of fouling, it still may be questionable if this is truly significant considering the margin of analytical errors.

To finalize discussion in this section, an explanation to relate membrane characterisation with rejection results is described. Based on the results of contact angles shown in Fig. 4.4, no apparent influence of an increase in hydrophobicity of the membranes was found to facilitate or diminish transport of compounds through the membrane. However, an increase in magnesium sulphate salt rejection from ~97% to ~98% (Fig. 4.5) may explain increased rejections by highly NOM fouled membranes (Fig. 4.15).

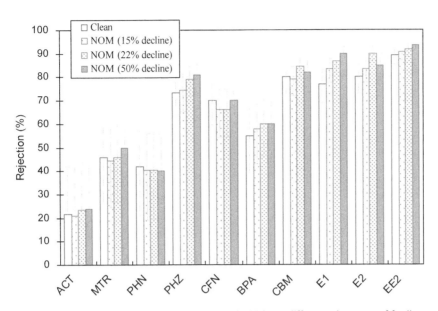

Fig. 4.15: Neutral compound rejections (NF-200) at different degrees of fouling

4.3.3   Rejection of ionic compounds by NF-200

The group of ionic compounds was expected to have electrostatic repulsion interactions with the membrane due to the negative charge of the membrane surface (Table 4.7) and the negative charge of the compounds. The results (Fig. 4.16) confirmed the described rejection mechanism; compounds were well rejected independent of molecular weight due to electrostatic repulsion. There is, however, a compound that presented lower rejection than the others, i.e., sulphamethoxazole (SFM).

As explained in the previous section regarding prediction capability, MW alone is not always sufficient in defining an observed rejection. Geometric size of the compounds can be more helpful with that. According to Figures 4.17, 4.18 and 4.19, SFM presented lower rejections due to minor equivalent width, length and effective diameter compared to ibuprofen (IBF). Ionic compounds showed increased rejections by membranes with 50% flux decline after NOM fouling (2–6%). Alginate fouling (22% flux decline) slightly increased rejections (0–4%). Dextran fouling (12% flux decline) and NOM fouling (15% flux decline) did not significantly increase rejections (0–1%).

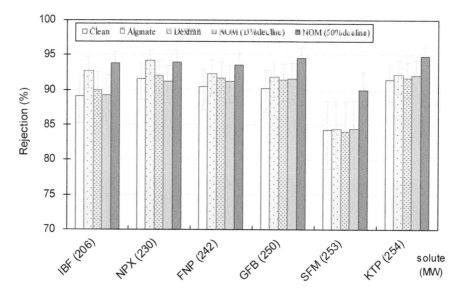

Fig. 4.16: Ionic compound rejections (MW) by NF-200 membranes

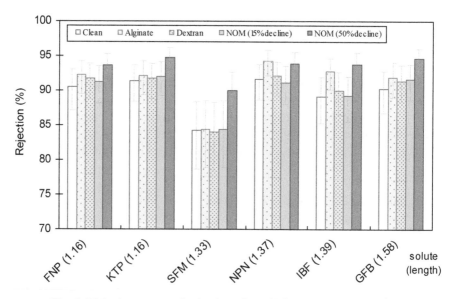

Fig. 4.17: Ionic compound rejections (length) by NF-200 membranes

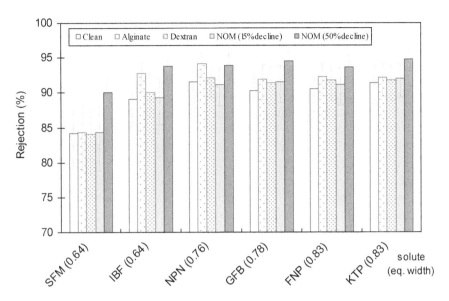

Fig. 4.18: Ionic compound rejections (eq. width) by NF-200 membranes

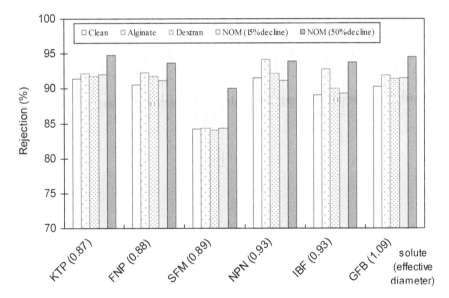

Fig. 4.19: Ionic compound rejections (effective diameter) by NF-200 membranes

4.3.4 Rejection of neutral compounds by NF-90

The main difference between NF-200 and NF-90 membranes is the MWCO. According to the manufacturer of the membranes, NF-200 and NF-90 have MWCOs of 300 and 200Da, respectively. Rejection of neutral compounds by NF-200 was related to size and hydrophobic interactions between solutes and the membrane. Rejection of neutral compounds by NF-90 membranes is shown in Fig. 4.20; it clearly can be recognized that the lower MWCO of NF-90 favoured the increase in rejection of compounds with MW > 200 Da. However, the low molecular weight compound (1,4-dioxane, DIX) passed through the membrane during filtration experiments. Figures 4.21, 4.22 and 4.23 give more information about the relationship between rejection and size (length, equivalent width and effective diameter) of the compounds. DIX was recognized as a compound with the lowest length and effective diameter; hence, size interactions between the membrane and solute probably resulted in low removal. It appears that for NF-90 membranes the effect of hydrophobicity of solutes (Fig. 4.24) and dipole moment (Fig. 4.25) have no impact in determining the increase or decrease in rejection. However, it may happen that a new compound with an adequate combination of size and hydrophobicity can partition through the membrane, as was previously elucidated for BPA with NF-200 membranes.

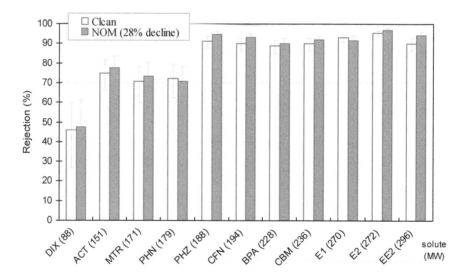

Fig. 4.20: Neutral compound rejections (MW) by NF-90 membranes

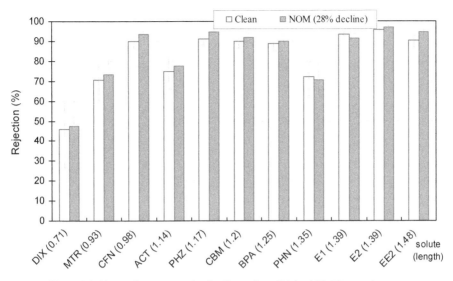

Fig. 4.21: Neutral compound rejections (length) by NF-90 membranes

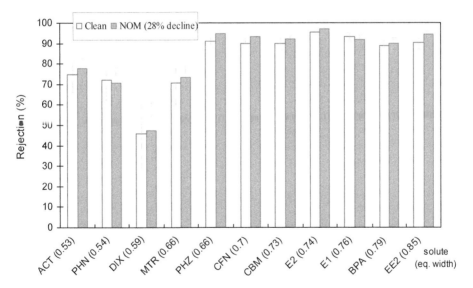

Fig. 4.22: Neutral compound rejections (eq. width) by NF-90 membranes

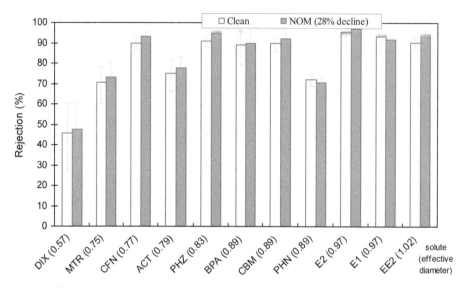

Fig. 4.23: Neutral compound rejections (effective diameter) by NF-90 membranes

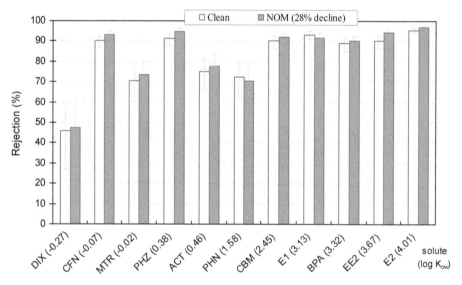

Fig. 4.24: Neutral compound rejections (log K_{ow}) by NF-90 membranes

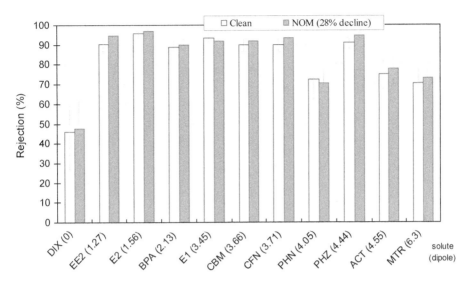

Fig. 4.25: Neutral compound rejections (dipole) by NF-90 membranes

4.3.5   Rejection of ionic compounds by NF-90

Rejection of negatively charged ionic compounds by NF-90 membranes is shown in Fig. 4.26. Rejection due to the effects of electrostatic interactions between the membrane and compound appear to have been been augmented by size exclusion mechanisms. Contrary to what was observed for NF-200 membranes, where size and possibly charge effects contributed to lower rejection of SFM, for the NF-90 membrane it can be hypothesized that SFM was mainly removed by size exclusion rather than by electrostatic repulsion. Fouling of NF-90 membranes with NOM did not present significant changes in the rejection of ionic compounds.

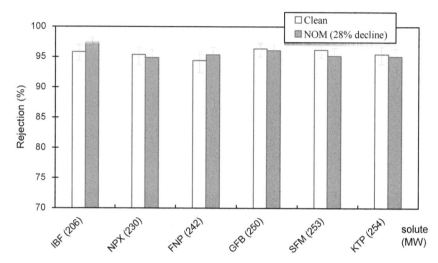

Fig. 4.26: Ionic compound rejections (MW) by NF-90 membranes

4.3.6   Rejection by NF-200 using different feed waters

To conclude this chapter, a comparison of rejections obtained using different feed waters (Table 4.9) is presented in this section. Experiments were carried out only for NF-200 membranes; the results are shown in Fig. 4.27 and Fig. 4.28. Rejection increased (6–46%) after direct fouling with NOM (35% flux decline) for all compounds, thus it can be hypothesized that the presence of humics, polysaccharides and divalent ions (Ca^{2+}, Mg^{2+}) in the feed water created a more compact fouling cake layer that acted as an additional (second membrane) barrier to remove compounds by size exclusion. The rejection of ionic compounds also increased; the cake layer that formed during direct filtration was more compact and may have removed ionic compounds by electrostatic repulsion and size exclusion. Li and Elimelech (2004) confirmed that divalent calcium ions greatly enhance NOM fouling by complexation and subsequent formation of intermolecular bridges among organic foulants molecules. In adittion, the study of Yangali (2005) demonstrated that colloidal NOM (polysaccharides) with high calcium concentration was the main cause of non-backwashable fouling (compact fouling layer) in UF membranes. The fouling of colloidal NOM with 103 mg/L Ca^{2+} was more pronounced than fouling of colloidal NOM with 4 mg/l Ca^{2+}. Comerton et al. (2008) also observed increased rejection of EDCs and PhACs by NF membranes from natural waters than from Milli-Q water. They attributed this finding to calcium ions forming complexes with organic matter in the fouling layer. Coagulation experiments using an inorganic coagulant (poly-silicate-iron, PSI) demonstrated that 17α-ethynylestradiol (a hydrophobic neutral compound) do not partition onto the total organic carbon (TOC) containing NOM (see Figure 4.29). The study of Vieno et al.

(2006) also concluded that coagulation does not entirely remove micropollutants from lake water. Therefore, the effective barriers for removal of micropollutants were the compact fouling layer and the membrane.

Table 4.9: Comparison of feed waters and fouling of NF-200 membrane

Name (flux decline)	Feed water	Fouling
NOM (22% decline)	i) fouling with 1.2μm pre-filtered water; ii) synthetic water + compounds cocktail	Pre-fouling with NOM
NOM (50% decline)	i) fouling with 1.2μm pre-filtered water; ii) synthetic water + compounds cocktail	Pre-fouling with NOM
Direct filtration (35% decline)	Delft canal water (1.2μm pre-filtered) + compounds cocktail	Direct fouling with NOM

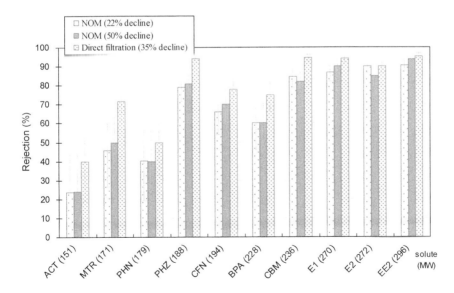

Fig. 4.27: Neutral compound rejections by NF-200 membranes

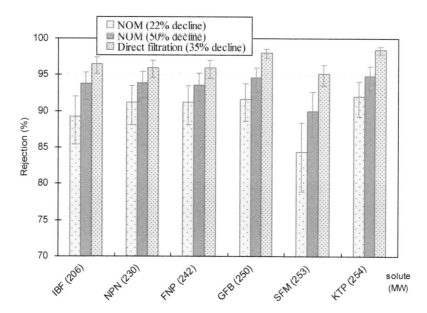

Fig. 4.28: Ionic compound rejections by NF-200 membranes

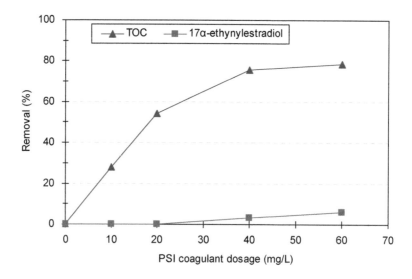

Fig. 4.29: Removal of TOC and 17α-ethynylestradiol after coagulation

4.4   Conclusions

○ Rejection of ionic compounds was high for NF-200 (>~85%) and NF-90 membranes (>~95%) and probably resulted from electrostatic repulsion; however, molecular size may influence rejection by steric hindrance as well.

○ Size represented by molecular length, effective diameter and equivalent width appeared to influence rejection of clean and fouled membranes. Rejection of neutral compounds by NF-200 was low to moderate (20–45%) when the equivalent width was smaller (<~0.6nm) than other equivalent widths of compounds with comparable or larger lengths. Nearly complete rejection of neutral compounds by NF-90 clean and fouled membranes was achieved, except for small size (low molecular weight) organic compounds.

○ Hydrophobicity effects represented by log K_{ow} or log D appeared to be important for compounds with log K_{ow} or log D values greater than 3 (negative effect). The hydrophobic effects after membrane saturation may favour solute partitioning if the size of the molecule allows partitioning (diffusion) through the membrane pore size. This effect was noticed only for BPA with NF-200 membranes.

○ The combined effect of size (molecular length, equivalent width or alternatively effective diameter) and log K_{ow} (or log D) resulted in greater rejection by NF-200 membranes, and nearly complete rejection by NF-90 membranes.

○ It appeared that high dipole moments decreased compound rejections; however, for the identified compounds, it was observed that their small molecular size appeared to be more important than their high dipole moments.

○ At a fouling degree equal to or less than 22% flux decline, mixed trends of increased or decreased (±9%) rejection were observed for neutral compounds by membranes fouled with sodium alginate, dextran and NOM. However, NOM fouling (up to 50% flux decline) appeared to increase rejection up to 14%, except for some compounds (caffeine, phenacetine).

○ It appeared that ionic compounds showed increased rejection by membranes after 50% flux decline was achieved with NOM fouling (2–6%). Alginate fouling (22% flux decline) slightly increased rejection (0–4%). Dextran fouling (12% flux decline) and NOM fouling (15% flux decline) had little effect on rejection (0–1%).

o   Direct filtration of feed water with NOM (35% flux decline) showed consistently increased rejection for all compounds (6–46%) mainly due to a dense compact layer of fouling. However, considering the particular use of one type of feed surface water, a general conclusion cannot be made for other types of feed water containing different compositions of NOM and divalent cations such as calcium and magnesium.

o   It was observed that a hydrophobic neutral compound (17α-ethynylestradiol) did not partition onto NOM.

o   Fouling of the membranes increased contact angle (according to the sessile drop method) indicating increased hydrophobicity. However, this fact did not correspond to the types of foulants used; specifically for dextran and alginate classified as neutral and hydrophilic, respectively. Apparently, hydrophobicity increased due to a compact layer of fouling and not due to the hydrophilicity/hydrophobicity of foulant used.

Chapter 5

A quantitative structure-activity relationship (QSAR) approach for modelling and prediction of rejections of organic solutes by NF membranes

Based on parts of:
- Modeling of RO/NF membrane rejections of PhACs and organic compounds: a statistical analysis, Drinking Water Engineering and Science, 1 (2008) 7–15.
- A QSAR Model for Predicting Rejection of Emerging Contaminants (pharmaceuticals, endocrine disruptors) by Nanofiltration Membranes, Water Research, (2009) In press, doi:10.1016/j.watres.2009.06.054
- Applications of quantitative structure-activity relationships for rejection of organic solutes by nanofiltration membranes, TECHNEAU Conference, Safe Drinking Water from Source to Tap State-of-the-Art and Perspectives, June 17-19, 2009, Maastricht, the Netherlands.

5.1   Introduction

Nanofiltration and reverse osmosis are technologies that provide medium to high rejections of organic compounds, present as emerging contaminants in water (Schafer et al., 2003; Kimura et al., 2003b). The presence of emerging contaminants has been identified in surface water bodies, sewage treatment plant effluents, and stages of drinking water treatment plants, and even at trace-levels in finished drinking water (Kolpin et al., 2002; Heberer, 2002; Castiglioni et al., 2006). The possible effects on aquatic organisms and human health, associated with the consumption of water containing low concentrations of single compounds, have been presented in toxicology studies (Escher et al., 2005; Pomati et al., 2006; Vosges et al., 2008). The studies demonstrate that researchers do not yet understand the exact risks from decades of persistent exposure to a myriad and random combination (of low levels) of pharmaceuticals, EDCs, and other organic contaminants; hence, the long-term effects of consumption of water containing low concentrations of contaminants will remain an unanswered question for the foreseeable future. Meanwhile, water treatment facilities are implementing monitoring programs, research organisations dealing with water reuse have published reports, and studies have addressed the topic (Drewes et al., 2006; Verliefde et al., 2007).

An important aspect of dealing with the problem has been the identification of compound physicochemical properties and membrane characteristics to explain transport, adsorption and removal of contaminants by different mechanisms, explicitly by size/steric exclusion, hydrophobic adsorption and partitioning, and electrostatic repulsion (Kiso et al., 2001b; Ozaki and Li, 2002; Van der Bruggen and Vandecasteele, 2002; Schafer et al., 2003; Kimura et al., 2003b; Nghiem et al., 2004; Bellona and Drewes, 2005; Xu et al., 2005). A number of articles have proposed a mechanistic understanding of the interaction between membranes and organic compounds; others have tried to apply complex models to model rejection (Cornelissen et al., 2005; Kim et al., 2007; Verliefde et al., 2008). However, there have been few models to "predict" the rejection of compounds.

To overcome this situation, our objective was to create a general quantitative structure-activity relationship (QSAR) model to predict rejection based on an integral approach that considers membrane characteristics, filtration operating conditions and physicochemical compound properties. A QSAR is a method that relates an activity of a set of compounds quantitatively to chemical descriptors (structure or property) of those compounds (Sawyer et al., 2003). QSAR's objective is predicting but maintaining a relationship to mechanistic interpretation.

Applications of QSAR for the development of models to find relationships between membranes and organic compounds have been presented in journals related to drug discovery and medicinal chemistry for analysis of permeability of membranes to organic compounds (Ren et al., 1996; Fujikawa et al., 2007). The study of reverse osmosis membranes has also experienced the application of QSAR principles. Campbell et al. (1999) performed a QSAR analysis of surfactants influencing the attachment of a mycobacterium to cellulose and aromatic polyamide reverse osmosis membranes. Their objective was to understand the relationship between surfactant molecular properties and activity on the membrane surface that inhibits bacterial attachment to the membrane to reduce biofilm formation and to increase permeate production.

The present study uses the concept of QSAR analysis to quantify an activity, compound rejection by a membrane, in terms of organic compound physicochemical properties, membrane characteristics (salt rejection, pure water permeability, molecular weight cut-off, charge, hydrophobicity) and operating conditions (pressure, flux, cross flow velocity, back diffusion mass transfer coefficient, recovery). In this work a QSAR model was constructed with internal experimental data used for training. The model was internally validated using measures of goodness of fit and prediction. Subsequently, after identification of a relationship in the form of an equation, estimations of rejection for an external dataset for different compounds and membranes were used to externally validate the model. Similarly, rejections of more emerging organic contaminants can be predicted in advance, before nanofiltration or reverse osmosis applications. Nevertheless, the QSAR model is applicable over the range of boundary experimental conditions that will be defined in the experimental section.

5.2   Experimental

The experimental aspect related to the equipment setup, compounds and membranes, and the analytical aspects that correspond to this chapter have been described in Section 3.3 of Chapter 3.

Table 5.1 presents the list of compounds present during filtration experiments. For hydrophobicity determination, log K_{ow} and log D were used; log K_{ow} is the octanol-water partition coefficient and log D is the ratio of the equilibrium concentrations of all species (unionised and ionised) of a molecule in octanol to the same species in the water phase. Log D values were calculated by ADME/Tox web software. Solubility and log K_{ow} values were obtained from the SRC Physprop experimental database. The dipole moment was calculated by Chem3D Ultra 7.0, Cambridgesoft. Size descriptors included molar volume (MV), molecular length, molecular width, molecular depth, equivalent molecular width and effective diameter. Molar volumes were calculated using the program ACD/ChemSketch Properties Batch, ACD/Labs. Molecular Modeling Pro, ChemSW, was used to compute size descriptors after optimization geometry of a molecule from the interaction of conformational analysis and energy minimization with a semi-empiric method MOPAC-PM3. Based on ionic speciation of compounds at pH 7 and log K_{ow} values, the compounds were classified as hydrophilic neutral, hydrophilic ionic, hydrophobic ionic and hydrophobic neutral.

Table 5.1: List of compounds used to generate internal data

Class.* / Name	MW (g/mol)	log K_{ow}[b]	log D^a (pH 7)	Dipole[c] (debye)	MV^d (cm³/mol)	Mol. length[e] (nm)	Mol. width[e] (nm)	Mol. depth[e] (nm)	Equiv. width[e] (nm)	Eff. diam.[e] (nm)
HL-neutral										
Acetaminophen	151	0.46	0.23	4.55	120.90	1.14	0.68	0.42	0.53	0.79
Phenacetine	179	1.58	1.68	4.05	163.00	1.35	0.69	0.42	0.54	0.89
Caffeine	194	-0.07	-0.45	3.71	133.30	0.98	0.87	0.56	0.70	0.77
Metronidazole	171	-0.02	-0.27	6.30	117.80	0.93	0.90	0.48	0.66	0.75
Phenazone	188	0.38	0.54	4.44	162.70	1.17	0.78	0.56	0.66	0.83
HL-ionic										
Sulphamethoxazole	253	0.89	-0.45	7.34	173.10	1.33	0.71	0.58	0.64	0.89
HB-ionic										
Naproxen	230	3.18	0.34	2.55	192.20	1.37	0.78	0.75	0.76	0.93
Ibuprofen	206	3.97	0.77	4.95	200.30	1.39	0.73	0.55	0.64	0.93
HB-neutral										
Carbamazepine	236	2.45	2.58	3.66	186.50	1.20	0.92	0.58	0.73	0.89
Atrazine	216	2.61	2.52	3.43	169.80	1.26	1.00	0.55	0.74	0.95
17 β-estradiol	272	4.01	3.94	1.56	232.60	1.39	0.85	0.65	0.74	0.97
Estrone	270	3.13	3.46	3.45	232.10	1.39	0.85	0.67	0.76	0.97
Nonylphenol	220	5.71	5.88	1.02	236.20	1.79	0.75	0.59	0.66	1.13
Bisphenol A	228	3.32	3.86	2.13	199.50	1.25	0.83	0.75	0.79	0.89

a. ADME/Tox Web Software
b. Experimental database: SRC PhysProp Database
c. Chem3D Ultra 7.0
d. ACD/ChemSketch Properties Batch
e. Molecular Modeling Pro
* HL = Hydrophilic, HB = Hydrophobic, hydrophobic if log K_{ow} > 2

The experiments produced a total internal dataset of 106 rejection cases; the dataset is presented as Appendix D. The boundary experimental conditions of the internal dataset are presented in Table 5.2. The internal dataset was used to develop the model. An external dataset that gathered three different datasets was used for validation of the model. The external dataset is presented as Appendix E. Experimental conditions for the first part of the external dataset can be obtained from a previous publication (Kim et al., 2007). Experimental conditions for the second and third part of the external dataset can also be found in a previous publication (Verliefde et al., 2008); the data correspond to filtration experiments using synthetic water solutions.

Table 5.2: Data range of membrane characteristics, operating conditions and rejections

Variable	Units	Min. value	Max. Value
Molecular weight cut-off (MWCO)	Da	200	300
Pure water permeability (PWP)	L/m²-day-kPa	0.86	2.23
Salt rejection (SR)*	-	0.96	0.98
Zeta potential (ZP, pH 7 & 10mM KCl)	mV	-48.04	-10.78
Contact angle (CA)	°	39.3	58.0
Pressure (P)	kPa	276	483
Cross-flow velocity (v)	cm/s	3	7.6
Back diffusion mass transfer coefficient (k)	cm/s	7.5E-04	1.3E-03
Flux (J)	L/m²-h	19	24
Hydrodynamic ratio (J_0/k)	-	1	2
Recovery	%	3	8
Rejection	%	17.7	99.0

* 2000mg/L $MgSO_4$, 25°C, recovery 15%, pressure 1034kPa, pH 8.

5.3   Results and discussion

5.3.1   QSAR methodology

A flow chart of the methodology used to build the QSAR model is illustrated in Fig. 5.1. The procedure to find a general QSAR equation to describe rejection was performed in four phases. The first phase was the organization of data from the experimental part. The data comprised 106 rejection cases. A total of 21 initial variables were used. The variables, considered as compound descriptors, were molecular weight (MW), solubility, log K_{ow}, log D, dipole moment, molar volume, molecular length, molecular width, molecular depth and equivalent width; variables describing membrane characteristics were molecular weight cut-off (MWCO), pure water permeability (PWP), magnesium sulphate salt rejection (SR), charge of the membrane as zeta potential (ZP), and hydrophobicity as contact angle (CA). The variables describing operating conditions were operating pressure (P), cross-flow velocity (v), back diffusion mass transfer coefficient (k), flux (J), ratio of pure water permeation flux J_0 and back diffusion mass transfer coefficient (J_0/k) and recovery. The range of values for membrane characteristics, operating conditions and rejections was presented in Table 5.2. The second phase was dedicated to the process of reducing variables using a correlation matrix and factor analysis with principal component analysis. The third phase corresponded to the regression analysis. In the third phase three methodologies were implemented: the first was a principal component analysis (PCA) with sequential application of multiple linear regression (MLR). The second method was the use of partial least squares (PLS) regression and MLR; and, the third method was the direct use of MLR. The last phase was the validation process. The model was internally validated using measures of goodness of fit (regression coefficients) and prediction (leave-one-out cross-validation); Section 5.3.5 provides details about the validation process. External validation of the general QSAR model was implemented by predicting rejections for an external dataset of experiments performed with different compounds and membranes, and with comparable operating conditions. PCA and PLS were performed using the research and statistical package SPSS Statistics 16.0. Leave-one-out cross-validations of the models were performed with MobyDigs (Talete, Milano, Italy).

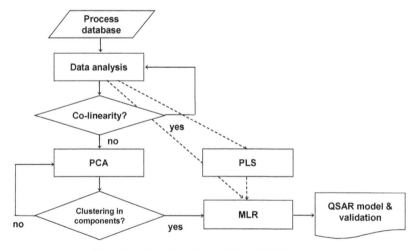

Fig. 5.1: Flow chart for building QSAR model

5.3.2   Variables reduction with PCA and QSAR model

The correlation matrix of the initial 21 variables (presented in Appendix F) was scrutinized in order to obtain a 'non-positive definite' matrix as a requisite of PCA. A matrix is called "non-positive definite" when there are both positive and negative eigenvalues. In the case of symmetric matrices, such as a correlation matrix, positive definiteness will only hold if the matrix and every principal sub-matrix have a positive determinant. A non-positive definite input matrix may signal a perfect linear dependency of one variable on another, known as co-linearity. This was the case for MWCO and salt rejection (SR) that had perfect linear correlation ($R^2=1$). Therefore, the application of PCA, considering independently one variable or the other, will give the same results of variable reduction and number of components. In other words, MWCO will not be excluded with PCA; the variable will be separated in advance and the results obtained for SR may be replaced by the variable MWCO, or vice versa.

Once an appropriate matrix was defined, the variables were analyzed in terms of how significant their correlations with rejection were; those correlations are also shown as an additional row and column of the 21×21 matrix. Rejection is only a reference variable to evaluate correlation with the rest of the variables. After a sequential implementation of PCA, three components were extracted; they defined the initial database of 21 variables with 11 variables describing three relationships, namely membrane/operating-conditions (component 1: flux, pure water permeability, salt rejection, zeta potential, mass transfer coefficient, cross-flow velocity), hydrophobicity/size (component 2: length, log K_{ow}, log D) and size (component 3: equivalent width, depth). Table 5.3 shows the loadings of variables per component.

Table 5.4 shows the contribution of each component in explaining the total variance of the variables. The final three rotated components explained 89.3% of the total variance, and the first two components explained 72.1%. It is important to mention that these results were produced for the internal dataset.

Table 5.3: Rotated component matrix

	Component		
	1	2	3
J	.979		
PWP	.967		
SR	.949		
ZP	-.936		
k	.936		
v	.880		
length		.951	
log K_{ow}		.930	
log D		.867	
eq. width			.972
depth			.910

Table 5.4: Total variance explained by components

Component	Initial Eigenvalues			Rotation Sums of Squared Loadings		
	Total	% of Variance	Cumulative %	Total	% of Variance	Cumulative %
1	5.331	48.466	48.466	5.323	48.388	48.388
2	3.084	28.035	76.501	2.609	23.718	72.107
3	1.411	12.824	89.326	1.894	17.219	89.326
4	.526	4.783	94.108			
5	.357	3.248	97.357			
6	.099	.902	98.259			
7	.089	.813	99.072			
8	.047	.431	99.503			
9	.041	.369	99.872			
10	.012	.111	99.983			
11	.002	.017	100.000			

The next step was the implementation of multiple linear regression (MLR) using the new set of variables. The use of MLR after PCA provides the advantage of a more simplified modelling approach. Moreover, the analysis of data before MLR may help to identify variables that are similar in response, which was the case for SR and MWCO. The dependent variable for all regression analyses was rejection. Two methods of linear regression were used; the first method is the enter (forced) method, which performs a regression with the contribution of all variables entered to model the dependent variable. The second method is stepwise regression, which is a more sophisticated method. Each variable is entered in sequence and its contribution is assessed according to an F-test. In the present study, an F-test with a statistical significance >0.10 implied removal of the variable, and an F-test with a significance <0.05 implied the entry of the variable in the model. If adding the variable contributes to the model then it is retained, but all other variables in the model are then re-tested to see if they are still contributing to the success of the model, otherwise an elimination process is carried out to remove those variables that are no longer judged to improve the model. Therefore, the method should ensure that the model contains a set of appropriate predictor variables.

Considering the previous explanation, the variables log K_{ow}, log D, length, depth, eqwidth, PWP, SR, ZP, v, k, J and the 106 rejection cases were used to model rejection. The final regression resulted in variables SR, eqwidth, log D, length and depth. The model resulted in a correlation coefficient R^2 of 0.75 with an F-test of 60.2, and a statistical significance of ~ 0%. Besides, all coefficients of the model variables showed very acceptable significances (< 0.001). Therefore, the QSAR linear equation model for rejection can be written as

$$rejection = \begin{array}{l} 265.150\,eqwidth - 117.356\,depth + 81.662\,length \\ -5.229\,log\,D + 1358.090\,SR - 1447.817 \end{array} \qquad (5.1)$$

where the units of the variables were specified in Table 5.1 and 5.2.

Eq. 5.1 can be mechanistically interpreted; rejection will increase by the effect of size, which is explained by the positive coefficients of length and equivalent width. The mechanism of steric hindrance due to size exclusion has been recognized as a main cause of rejection in many studies (Van der Bruggen et al., 1999; Kiso et al., 2001b; Ozaki and Li, 2002; Nghiem et al., 2005). By contrast, the negative coefficient associated with log D infers decreased rejection, which clearly states that the effect of hydrophobicity lessens rejection due to adsorption and subsequent partitioning mechanisms. Indeed, partitioning is a combined effect that is not only compound property dependent (size, hydrophobicity) but also is related to membrane characteristics. It is important to mention that log D is assuming the role of hydrophobicity for neutral and ionic compounds, compounds with high log D values will adsorb to the membrane and may partition up to saturation. Conversely, ionic compounds and very hydrophilic neutral compounds are represented by log D values that are very low or negative, thus will not adsorb, and actually may adsorb at greater values of log D (~ 2–3). Hydrophobicity influences rejection after adsorptive interactions with the membrane; this fact has been documented in some studies (Kiso et al., 2001b; Kimura et al., 2004). The role of depth in the equation will compensate size exclusion contributions of length and equivalent width in a final rejection. The equation also shows that salt rejection (SR) is a parameter incorporating steric/size hindrance and electrostatic repulsion effects related to the charge of the membrane and operating conditions. This effect may be related to cake-enhanced concentration polarization affecting the salt rejection of clean and fouled membranes (Hoek and Elimelech, 2003). Thus, SR ultimately serves as a comparison parameter between membranes of the same type (aromatic polyamide) but with minor differences in pore size, and possibly with differences in charge. The QSAR equation merges information about the interaction of membrane characteristics, filtration operating conditions and organic compound properties to predict rejections during nanofiltration. According to Eq. 5.1, the contact angle and zeta potential as measurements of hydrophobicity/hydrophilicity and membrane surface charge, respectively, were not part of the equation and therefore did not contribute quantitatively to model rejection. However, size/steric hindrance effects related to salt rejection, and hydrophobicity of the solutes were part of the model's equation. In conclusion, rejection increased by size/steric hindrance effects; hydrophobicity, however, decreased rejection due to adsorption and partitioning mechanisms.

The results of PCA and the model cannot be generalized to experimental conditions outside the boundary experimental conditions. For instance, excessive changes of pH affect the ionic speciation of charged compounds, obviously pK_a values of solutes and the pH of feed waters will determine boundary conditions for applicability of the model. Changes in membrane properties such as charge and pore size due to swelling will also influence the model. Other considerations for application of the model are the type of membrane used (aromatic polyamide), fluxes, pressures and cross-flow velocities. Cellulose acetate or even different membrane compositions (other than aromatic polyamide) for NF-90 or NF-200 will influence the PCA and the model. Nonetheless, the approach is valid and can be generalized under certain conditions in up-scaling NF applications.

5.3.3   QSAR model after PLS regression and MLR

An introduction to partial least-squares (PLS) regression was presented in Sec. 2.6.3. The following variables were removed after PLS regression: MW, solubility, MV, MWCO, ZP, CA, P, v, J, J_0/k, recovery and width. Therefore, the final PLS model is defined by the variables log K_{ow}, log D, dipole, length, depth, equivalent width (eq. width), PWP, SR and k. The main advantage of PLS is its ability to handle co-linearity among the independent variables. In contrast, principal component analysis cannot control co-linearity. Another advantage of PLS regression is that the calculation process is simpler. However, PLS is used more as a predictive technique and not as an interpretive one, like MLR. Therefore, in this exploratory analysis, PLS regression serves as a variable selection process and as a prelude to implementation of MLR. Once again, as occurred with PCA, a reduced number of variables simplified the implementation of MLR. After applying stepwise regression to the reduced number of variables, the model result obtained was also Eq. 5.1.

5.3.4   QSAR model after MLR

To finalize the model development under various statistical application scenarios, the implementation of direct multiple linear regression was performed. The R^2 is calculated for all possible subset models. Using this technique, the model with the largest R^2 is declared the best linear model. However, this technique has some disadvantages. First, the R^2 increases with each variable included in the model. Therefore, this approach encourages including all variables in the best model, although some variables may not significantly contribute to the model. This approach also contradicts the principal of parsimony that encourages as few parameters in a model as possible. Thus, the application of MLR without prior data analysis is a possibility when the number of variables is limited. Another disadvantage experienced during the present study was that MLR was not able to distinguish co-linearity between variables. This was the case for variables MWCO and SR; MWCO can replace the role of SR in Eq. 5.1. The new equation resulted in an R^2 of 0.75; the F-test was 52.5 with a significance of ~0%. The equation was

$$rejection = \frac{265.150\,eqwidth - 117.356\,depth + 81.662\,length}{-5.229\,log\,D - 0.272\,MWCO - 62.565}$$

$$(5.2)$$

It is important to mention that Eq. 5.2 would have been equally defined in Section 5.3.2 if the variable selected before implementing PCA was MWCO and not SR as previously explained; therefore it is not surprising that Eq. 5.1 and 5.2 only show differences in two coefficients. However, considering practical or operational factors, it may be more difficult to determine changes in MWCO when fouling occurs; on the other hand, salt rejection tests are part of the monitoring practice. In addition, the variable of MWCO may be difficult to define when a range of MWCO exists for a membrane rather than an approximate single MWCO value if compared to salt rejection. A range of MWCO is not desirable for a membrane because this will greatly influence the rejection of solutes with sizes close to or in the range of MWCO.

It may be argued that a simple stepwise MLR will produce the same results as a PCA or PLS followed by MLR; however, the application of MLR without PCA or PLS has the disadvantage of removing or adding appropriate /inappropriate variables during the iterative stepwise MLR process. This process is only dependent on the fulfilment of a statistical condition; it may even happen that a different combination of variables defines a good equation. Therefore, the advantage of PCA or PLS is that only important variables are considered, and only those will be part of the final MLR implementation.

5.3.5 Validation of linear and non-linear QSAR model

Actual (measured) rejection values (106 rejection cases) versus modelled (fitted) rejections of the data used to generate the model are shown in Fig. 5.2 and Fig. 5.3, a 95% confidence interval shows that very few modelled rejections (outliers corresponding to the NF-200 membrane) were outside of that interval. Besides the good fit of a model, it is necessary to assess its predictive power, i.e. its robustness (Eriksson et al., 2003). The R^2 is the most widely used measure of the ability of a QSAR model to reproduce the internal data in the training (goodness of fit), but does not explain its robustness and predictive power.

One technique to evaluate prediction is the leave-one-out cross-validation technique, in which one case at a time is iteratively held-out from the training set and the rest is used for model development; the excluded case is predicted by the developed model (Gramatica, 2007). According to Gramatica, the predictive power of a model may be estimated by the goodness of prediction parameter Q^2 leave-one-out (1-PRESS/TSS, where PRESS is the predictive error sum of squares and TSS is the total sum of squares). In general, a $Q^2 > 0.5$ is regarded as good and $Q^2 > 0.9$ as excellent (Eriksson et al., 2003). For the developed QSAR models, the model with SR

(Eq. 5.1) presented a Q² leave-one-out of 0.72, and the model with MWCO (Eq. 5.2) presented a Q² leave-one-out of 0.72.

After internal cross-validation it was demonstrated that Equations 5.1 and 5.2 were valid to model rejection, however, an adjustment must be made to the equation before using it to compare measured vs. predicted rejections for external databases. This adjustment was necessary to overcome the mathematical structure of the equation. Using a physical interpretation, it was evident that size parameters referring to the variables length and equivalent width may be large enough to cause rejection predictions over 100%, which can be explained after observing positive coefficients for equivalent width and length. This situation may also be detrimental for rejection predictions of ionic compounds of medium to large size (0.6–1.2 nm as equivalent width) that are generally rejected due to electrostatic repulsion and less steric hindrance. Therefore, Eq. 5.1 and 5.2 can be transformed to the following conditional equation

$$rejection = \begin{cases} 100 & if\ QSAR\ model \geq 100 \\ QSAR\ model \end{cases}$$

(5.3)

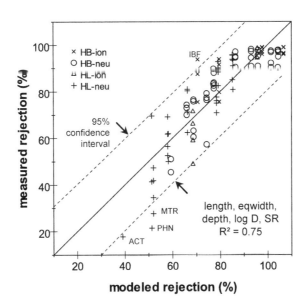

Fig. 5.2: QSAR model of experimental internal database with SR

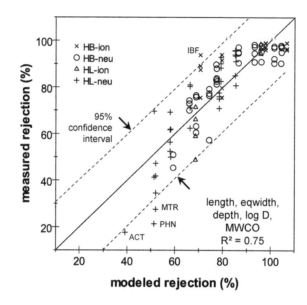

Fig. 5.3: QSAR model of experimental internal database with MWCO

In order to improve or, in any case, investigate the existence of non-linear relations and to obtain an asymptotic equation limiting rejections to 100% and set aside the conditional Eq. 5.3, a non-linear model was developed for SR. The equation for the non-linear model is

$$rejection = \frac{100}{1+e^{-\alpha}} \qquad (5.4)$$

where α is equal to

$$\alpha = 6.283 length + 19.377 eqwidth + 108.337 SR - 0.443 \log D - 8.112 depth - 119.146$$

Fig. 5.4 shows modelling rejection results of the non-linear model of Eq. 5.4 with a confidence interval of 95%, and four identified outliers corresponding to the NF-200 membrane.

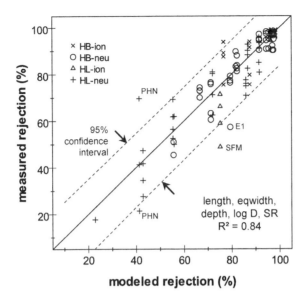

Fig. 5.4: Non-linear QSAR model of experimental internal database (Eq. 5.4)

It was mentioned that the role of depth is to compensate size exclusion contributions of length and equivalent width in a final rejection. This explanation may be better understood after replacing length and depth by a more representative size variable. The hypothesis is that the effective diameter can be an appropiate replacement because it represents an average projection of how the solute approachs the membrane (Sec, 2.3.1). Therefore, the equation derived after considering the effective diameter as variable is

$$rejection = \frac{100}{1 + e^{-\beta}} \qquad (5.5)$$

where β is equal to

$$\beta = 11.372\,effdiameter + 6.795\,eqwidth + 107.305\,SR - 0.399\log D - 116.496$$

For Eq. 5.5, Fig. 5.5 shows modelling rejection results of the non-linear model with a confidence interval of 95%, and four identified outliers corresponding to the NF-200 membrane.

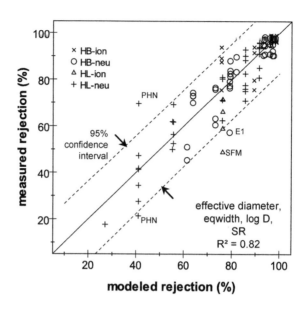

Fig. 5.5: Non-linear QSAR model of experimental internal database (Eq. 5.5)

An external dataset (that gathered three different datasets) was selected for external validation of the QSAR model. The first part of the external dataset corresponds to the membrane Filmtec NF-90. The second part corresponds to the NF membrane Trisep TS-80, and the third part corresponds to the NF membrane Desal HL (GE Osmonics). It is worthwhile to mention that the second and third parts of the external dataset were generated using spiral wound membrane elements instead of flat sheet membranes, but under comparable experimental conditions.

Fig. 5.6 illustrates the results of measured rejections vs. predicted rejections after calculations with Eq. 5.3, for the QSAR model with Eq. 5.1 (SR). Fig. 5.7 shows results of measured rejections vs. predicted rejections after calculations with Eq. 5.3, for the QSAR model with Eq. 5.2 (MWCO). Fig. 5.8 illustrates the results of measured rejections vs. predicted rejections after calculations with Eq. 5.4, for the non-linear QSAR model (length, equivalent width, depth, log D, SR). Finally, Fig. 5.9 illustrates the results of measured rejections vs. predicted rejections after calculations with Eq. 5.5, for the non-linear QSAR model (effective diameter, equivalent width, log D, SR).

In order to determine which of the models was the best model for prediction; an error parameter was determined. The standard deviation of error (STDE), measured as a percentage, provides an unbiased measure of the model's performance compared to the regression coefficient (R^2) that only measures the regression response between predicted and measured rejections. The R^2 for predictions of the QSAR model with SR was 0.88 with an STDE of 9% (Fig. 5.6). The R^2 for predictions of the QSAR model with MWCO was 0.85 with an STDE of 11% (Fig. 5.7). The best models were the non-linear QSAR models, with an R^2 of 0.93 and an STDE of 7% (Fig. 5.8) for Eq. 5.4; and, with an R^2 of 0.94 and an STDE of 6% (Fig. 5.9) for Eq. 5.5.

It is important to mention that organic compounds with rejections of less than 40% have been identified (labelled) in the figures. Chloroform (CF), perchloroethene (PCE), carbon tetrachloride (CT), 2-methoxyethanol (MET), ethanol (ETH), 2-ethoxyethanol (EET), 2-(1H)-Quinoline (QNL), glycerol (GLY) and N-Nitrosodimethylamine (NDMA) were low molecular weight compounds used in external experiments. It can be observed that the model has been able to extrapolate rejections of small compounds not present during the developmental phase of the model. An improved model can be obtained when those compounds are included in the initial dataset that defines the model.

The main difference between the model results shown in Figs. 5.6 and 5.7 was that the model with MWCO (Fig. 5.7) showed a lower R^2 (0.85) and STDE (11%) than the linear model with SR (Fig. 5.6, $R^2 = 0.88$) and STDE (9%), meaning that the latter had a better goodness of fit and prediction performance for external data. Moreover, the characterisation of magnesium sulphate salt rejection for a membrane may be preferred over MWCO, particularly for nanofiltration; besides, the effect of fouling on membranes can also be quantified by salt rejection experiments. The linear and non-linear QSAR models with SR were demonstrated to be acceptable for the external dataset of NF-90, Trisep TS-80 and Desal HL. Although the model can be valid with limitations related to boundary experimental conditions, its applicability and approach can be of value for the construction of a model with combined datasets organised in training and testing groups.

A bimodal trend appears to be present between modelling results obtained with SR and MWCO. This fact is interesting and reveals that both variables are indeed correlated as initially determined for the NF-90 and NF-200 membranes. It should also be noticed (Figs. 5.6 and 5.7) that NF-90 and Trisep TS-80 are membranes that present more support for the correlation between SR and MWCO.

It has been demonstrated that the effective diameter can be a good replacement of length and depth, with prediction results even better than the latter ones (Figs. 5.8 and 5.9). However, the effective diameter may not be the ultimate size descriptor because the modelling (predicting) equation still requires the contribution of the equivalent width to achieve good prediction results.

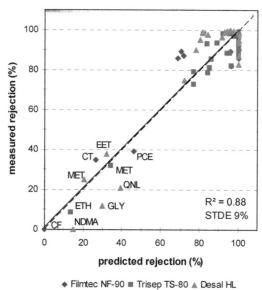

Fig. 5.6: Predictions of external database with linear QSAR (SR)

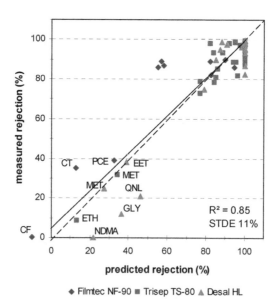

Fig. 5.7: Predictions of external database with linear QSAR (MWCO)

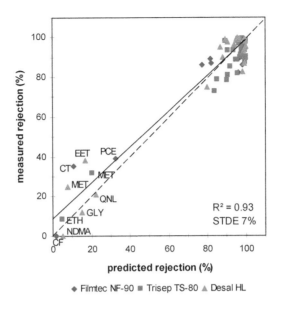

Fig. 5.8: Predictions of external database with non linear QSAR (Eq. 5.4)

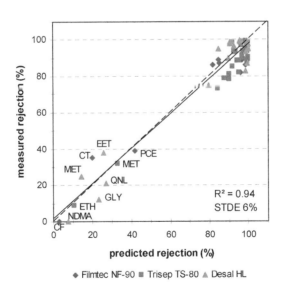

Fig. 5.9: Predictions of external database with non linear QSAR (Eq. 5.5)

5.4   Conclusions

o A general QSAR model equation can be developed to integrate
   information about the interaction of membrane characteristics,
   filtration operating conditions and solute properties to predict
   rejection of emerging contaminants during nanofiltration.

o The best QSAR models identified that the most important variables
   that may influence rejection of organic solutes were log D, salt
   rejection, equivalent width, effective diameter, depth and length.

o The non-linear QSAR model showed better performance than the
   linear QSAR model.

o The results suggest that molecular size descriptors play an important
   role in the non-linear QSAR model. The effective diameter and the
   equivalent width were identified as the best molecular size
   descriptors in membrane rejection.

o Rejection increased by size/steric hindrance effects; solute
   hydrophobicity decreased rejection due to adsorption and
   partitioning mechanisms.

o Salt rejection ($MgSO_4$) captured steric hindrance and electrostatic
   repulsion effects that were related to the membrane structure and
   operating conditions.

o The use of MWCO was acceptable for modelling purposes;
   however, NF membranes with a broad range of MWCO (pore size
   and distribution) may make it difficult to estimate rejection of
   contaminants, thus magnesium sulphate salt rejection may be more
   appropriate. Nonetheless, a good correlation between salt rejection
   and MWCO may exist.

Chapter 6

Data-driven modelling applying QSAR and ANN to predict rejection of neutral organic compounds by NF and RO membranes

Based on parts of:
- Artificial neural network models based on QSAR for predicting rejection of neutral organic compounds by polyamide nanofiltration and reverse osmosis membranes, Journal of Membrane Science, 342 (2009), 251-262.
- A QSAR (quantitative structure-activity relationship) approach for modelling and prediction of rejection of emerging contaminants by NF membranes. Desalination and Water Treatment. In Press.

6.1   Introduction

Until recently, the modelling of rejection by NF and RO membranes of charged organic compounds was limited mainly to negatively charged compounds, and rejection was explained by mechanisms of electrostatic repulsion due to the negative charge of the membrane (Kimura et al., 2003b; Bellona and Drewes, 2005; Yoon et al., 2006; Nghiem et al., 2006). Verliefde et al. (2008) contributed to a better understanding of rejection mechanisms for negatively and positively charged organic compounds; however, their rejection model is based on the determination of rejection of equivalent non-charged organic compounds. The rejection of charged compounds is highly dependent on the charge of the membrane. Verliefde et al. (2008) showed that if the membrane is negatively charged it will repel negatively charged compounds but will attract positively charged compounds, creating an effect of charge concentration polarization that will result in more transport and less rejection of positively charged compounds. On the other hand, rejection mechanisms of neutral compounds have been addressed in many publications. Kiso et al. (2000, 2001b) concluded that rejection of hydrophobic neutral compounds correlated with molecular width and size in addition to hydrophobicity. Ozaki and Li (2002) observed that rejection of neutral organic compounds by a low pressure RO membrane increased linearly with molecular weight and molecular width. Kimura et al. (2003b) as well as Berg et al. (1997) found that the rejection of neutral organic compounds by an NF and a low pressure RO was mainly influenced by the molecular size of the compounds. Regarding size and hydrophobicity, Schäfer et al. (2003) indicated that both size exclusion and adsorption are essential in maintaining high initial retention by NF membranes. Similarly, Nghiem et al. (2004) concluded that steroid hormone diffusion in the membrane polymeric matrix most likely depends on the size of the hormone molecule and hydrophobic interactions of the hormone with the membrane polymeric matrix. Other investigations have indicated that rejection of neutral compounds appeared to be notably influenced by molecular width and length (Chen et al., 2004; Xu et al., 2005; Agenson et al., 2003).

After an extensive literature review, Bellona et al. (2004) concluded that to be able to quantify the rejection of a certain organic solute by a particular membrane type it is important that physicochemical characteristics such as solute ionization potential (charged or neutral), solute hydrophobicity, and molecular size be considered as the most influential parameters responsible for rejection. Mechanistic models used to describe rejection of neutral solutes are often represented by one molecular size predictor describing steric hindrance factors within a membrane pore. However, considering only one size parameter (molecular weight, molar volume or equivalent width) in the prediction of the rejection of neutral organic solutes will most likely not be sufficient to represent solute size, as well as solute membrane interactions (Kiso et al., 1992; Van der Bruggen et al., 2000; Schafer et al., 2003; Kimura et al., 2003a; Bellona et al., 2004; Yoon et al., 2006).

In addition to physicochemical properties, membrane characteristics and operating conditions are needed to understand membrane rejection. It has been demonstrated that size exclusion plays an important role in membrane rejection. Schafer et al. (2005) observed that MWCO may not infer the true state of retention of solutes smaller than the MWCO since the dimensional parameters of the molecules are not taken into account in this case, and therefore retention of molecules with a smaller molecular weight but different molecular structures may differ. In another study by Kimura et al. (2004), it was demonstrated that MWCO cannot be used to predict the rejection of EDCs/PhACs by RO membranes since properties of standard compounds used for MWCO determination and those of EDCs/PhACs may be different.

It is evident that the use of MWCO raises problems when experimental or operating conditions are changed. For instance, changes in rejection when a membrane becomes fouled is a topic that has not been addressed in rejection prediction with current mechanistic models, and only qualitative explanations are available to explain the fouling effects on rejection (Xu et al., 2006; Lee et al., 2006; Li et al., 2007; Comerton et al., 2008). In an attempt to define a general model, some publications have indicated the complexity of mechanistic models to simulate organic compound rejection by NF and RO membranes; that complexity has led to a number of assumptions that simplify mechanistic model development (Agenson et al., 2003; Ben-David et al., 2006; Bowen and Mohammad, 1998; Cornelissen et al., 2005; Kim et al., 2007).

The work of Libotean et al. (2008) is an interesting approach to modelling that uses information of quantitative structure-property relationships that correlates organic solute rejection with membrane properties to build an artificial neural network model. In their article, predictive models for rejection of organic micropollutants by NF and RO membranes are developed using artificial neural networks based on information obtained from quantitative structure-activity relationship (QSAR) models. The main differences between the approach taken by Libotean et al. and the approach presented in this work are: the use of three fundamental descriptors (in this work) rather than 45 fundamental molecular descriptors, a different model development approach that will be explained in Sections 6.4, 6.5.1 and 6.5.2, and the use of internal experimental data and external data.

Fundamental molecular descriptors are calculated from physical or chemical properties, and from topological indices derived from connectivity and composition of a molecule; they are also known as non-empirical molecular descriptors. Non-fundamental descriptors are empirical descriptors, which are not derived from the composition or structure of a molecule; they can be obtained from non-empirical descriptors or from experimental data. An explanation of the reasons that motivated the selection of a limited number of fundamental descriptors is given in Section 6.3. Fundamental molecular descriptors used in this study were molecular weight, dipole moment and molar volume; non-fundamental geometrical descriptors that were used are molecular length, width, depth and equivalent width; and a non-fundamental hydrophobicity descriptor ($\log K_{ow}$) was used as well.

Only polyamide NF and RO membranes were investigated; the membrane characteristics considered were molecular weight cut-off, pure water permeability, salt rejection of magnesium sulphate, surface membrane charge (as zeta potential) and hydrophobicity (as contact angle). Additionally, operating conditions such as pressure and permeate flux were considered. The ANN models presented in this article were constructed using internal experimental data (77%) and external data (23%). The approach to develop the QSAR model was applied over a combination of variables related to solutes, membranes and operating conditions.

The ANN models were built extracting information from the QSAR model; only 60% of the total data was used to develop the ANN-QSAR based model, 20% of the data was used for validation and the remaining 20% was used for independent predictions. The potential advantages of the most promising proposed models are that i) they combine variables related to solutes and membranes, ii) internal experimental data and external data were used to generate the model, and iii) magnesium sulphate salt rejection was introduced as a lump parameter for modelling the rejection of contaminants by NF and RO membranes. The limitations of the selected models are that i) they apply only to neutral organic compounds, ii) they are only valid over the range of boundary experimental conditions of the database.

6.2   Experimental

As mentioned in the introduction, in order to expand the amount of data, internal experimental data were used in combination with external data from the literature. The internal data comprised 124 rejection cases of 36 organic compounds by 5 NF and 5 RO membranes. The experimental methodology for the internal data was described in Section 3.3 of Chapter 3 and in previous publications by Kim et al. (2005, 2007) and Verliefde et al. (2008).

In order to increase the number of solutes and membranes and therefore the number of rejection cases, additional data on rejections (37 cases) of organic solutes by 1 NF and 5 RO membranes were obtained from Van der Bruggen et al. (1999), Ozaki and Li (2002), Kimura et al. (2004) and Yoon and Lueptow (2005).

The list of the neutral organic compounds and physicochemical descriptors is presented in Table 6.1; the list shows compounds that were part of the internal data and references to literature from which other compounds or the same compounds were part of the external data. The list of membranes and experimental conditions corresponding to internal and external data is presented in Tables 6.2 and 6.3. The range of experimental conditions that defines the boundary experimental conditions of models is also presented in those tables. The experimental approach taken in the external referenced work was comparable to the experimental approach adopted by the authors in their own experimental work. Experiments were carried out in laboratory scale cross-flow units (90% of the data), except for one case, in which the experimental setup used a stirred cell dead-end unit (10% of the data). The effects of concentration polarization were considered to be negligible because the experiments used deionised water or ionic strength solutions of less than or equal to 10mM KCl; cross-flow velocities and the use of spacers (or stirring in one external experiment) also contributed to reducing the concentration polarization. Rejection was calculated using Equation 3.4 presented in Chapter 3.

Table 6.1: List of compounds and physicochemical properties used for ANN model

#	Name	ID	MW (g/mol)	log K_{ow}	Dipole (debye)	length (nm)	Eq. width (nm)	Ref.
1	17β-estradiol	BES	272.39	4.01	1.56	1.39	0.74	a, b
2	2-(1H)-Quinoline	QNL	145.16	1.26	3.38	1.00	0.52	a
3	2-(2-Butoxyethoxy)	BUE	162.22	0.56	1.41	1.61	0.53	e
4	2,4-Dichlorophenol	DCL	163.00	3.06	0.40	0.92	0.51	d
5	2-ethoxyethanol	EET	90.12	-0.32	0.41	1.00	0.53	a
6	2-methoxyethanol	MET	76.10	-0.77	0.25	0.87	0.52	a
7	4-Chlorophenol	CHP	128.56	2.39	1.48	0.92	0.48	d
8	Aminopyrine	APY	231.30	1.00	3.19	1.27	0.73	a
9	Antipyrine	ANP	188.23	0.38	5.42	1.17	0.66	a
10	Atrazine	ATZ	215.68	2.61	3.43	1.26	0.74	a
11	Bentazon	BTZ	240.28	2.34	1.29	1.18	0.76	a
12	Benzyl alcohol	BEA	108.14	1.10	1.46	0.90	0.53	c, d

Table continue in next page

Cont.

#	Name	ID	MW (g/mol)	log K_{ow}	Dipole (debye)	length (nm)	Eq. width (nm)	Ref.
13	Bisphenol A	BPA	228.29	3.32	2.13	1.25	0.79	a, b
14	Bromoform	BF	252.73	2.40	1.00	0.69	0.56	a
15	Caffeine	CFN	194.19	-0.07	3.71	0.98	0.70	a, b
16	Caprolactam	CAL	113.16	0.66	3.73	0.79	0.65	a, e
17	Carbamazepine	CBZ	236.27	2.45	3.66	1.20	0.73	a, b
18	Carbontetrachloride	CT	153.82	2.83	0.30	0.64	0.60	a
19	Chloroform	CF	119.38	1.97	1.12	0.53	0.42	a
20	Chlorotoluron	CTL	212.68	2.41	2.74	1.29	0.61	a
21	Cyclophosphamide	CPA	261.09	0.63	4.50	1.12	0.78	a
22	Diuron	DIU	233.10	2.68	1.12	1.31	0.56	a
23	Estrone	ESN	270.37	3.13	3.45	1.39	0.76	a
24	Ethanol	ETH	46.07	-0.31	1.55	0.64	0.52	a
25	Ethylene glycol	ETG	62.07	-1.36	0.00	0.76	0.52	d
26	Formaldehyde	FOM	30.03	0.35	2.16	0.47	0.38	e
27	Glucose	GLU	180.16	-3.24	2.23	0.94	0.75	a
28	Isopropanol	ISP	60.11	0.05	2.67	0.66	0.56	c, e
29	Isoproturon	IPT	206.29	2.87	2.40	1.42	0.66	a
30	Lindane	LID	290.83	3.72	1.00	0.91	0.78	a
31	Methacetin	MTC	165.19	1.03	2.20	1.28	0.52	a
32	Methanol	MTH	32.04	-0.77	1.62	0.53	0.44	c, e
33	Methylethylketon	MEK	72.12	0.29	2.74	0.72	0.59	c
34	Methylmetacrylate	MEM	100.13	1.38	1.81	0.88	0.53	c
35	Metobromuron	MBM	259.10	2.38	0.11	1.34	0.61	a
36	Metoxuron	MTX	228.68	1.64	3.03	1.29	0.70	a
37	Metronidazole	MTR	171.16	-0.02	6.30	0.93	0.66	a
38	Monolinuron	MLN	214.65	2.30	0.30	1.22	0.69	a
39	NAC standard	NAC	201.22	2.36	2.26	1.24	0.60	b
40	Nitrobenzene	NIB	123.11	1.85	5.26	0.85	0.48	c
41	Nonylphenol	NPL	220.35	5.71	1.02	1.79	0.66	a
42	Pentoxifylline	PFL	278.31	0.29	4.78	1.52	0.81	a
43	Perchloroethene	PCE	165.83	3.40	0.11	0.78	0.59	a
44	Phenacetine	PHN	179.22	1.58	4.05	1.35	0.54	a, b
45	Phenazone	PHZ	188.23	0.38	4.44	1.17	0.66	a
46	Primidone	PRI	218.25	0.91	2.82	1.10	0.76	a, b
47	Simazin	SMZ	201.66	2.18	0.01	1.37	0.64	a
48	Trichloroethene	TCE	131.39	2.29	0.95	0.78	0.49	a
49	Triethylene glycol	TEG	150.17	-1.75	0.00	1.47	0.52	d
50	Urea	URE	60.06	-2.11	-1.68	0.67	0.43	d, e

a. internal data; b. Kimura et al., 2004; c. Van der Bruggen et al., 1999; d. Ozaki and Lee, 2002; e. Yoon and Lueptow, 2004.

Table 6.2: List of membranes and experimental conditions

Product name	Type	Manufacturer	Experiment	Feed conc.	pH	Ref.
Internal						
NF-90	NF	Dow-Filmtec	cross-flow	6.5-100µg/L	7-8	a
NF-200	NF	Dow-Filmtec	cross-flow	6.5-100µg/L	7-8	a
XLE-440	LPRO	Dow-Filmtec	cross-flow	100 µg/L	8	a
LE-440	LPRO	Dow-Filmtec	cross-flow	100 µg/L	8	a
BW-440	RO	Dow-Filmtec	cross-flow	100 µg/L	8	a
RE-BLR	RO	Saehan	cross-flow	100 µg/L	8	a
NE-90	NF	Saehan	cross-flow	100 µg/L	8	a
UTC-70	LPRO	Toray	cross-flow	100 µg/L	8	a
TS-80	NF	Trisep	cross-flow	2 µg/L	7	a
Desal-HL	NF	GE Osmonics	cross-flow	2 µg/L	7	a
External						
ES-20	LPRO	Nitto Denko	cross-flow	10 mg/L	7	d
NF-70	NF	Dow-Filmtec	cross-flow	200-400mg/L	7	c
XLE-440	LPRO	Dow-Filmtec	cross-flow	100 µg/L	8	b
AK	LPRO	Desal-Osmonics	stirred cell	10 mg/L	7.5	e
ESPA	LPRO	Hydranautics	stirred cell	10 mg/L	7.5	e
ESNA	LPRO	Hydranautics	stirred cell	10 mg/L	7.5	e

a. Internal data.
b. Kimura et al., 2004.
c. Van der Bruggen et al., 1999.
d. Ozaki and Lee, 2002.
e. Yoon and Lueptow, 2004.
LPRO (low pressure reverse osmosis)

Table 6.3: Membrane characteristics and operating conditions

Membrane	Type	MWCO estimate (Da)	PWP (L/m²-day-kPa)	SR[a]	ZP[b] (mV)	CA (°)	Flux (L/m²-h)	Pressure (kPa)
Internal								
NF-90	NF	200	2.23	0.98	-27.00	59.8	22–24	280
NF-200	NF	300	1.01	0.96	-20.00	37.5	19–24	480
XLE-440	LPRO	150	0.92	0.98	-19.42	39.8	16	410
LE-440	LPRO	100	0.77	1.00	-23.02	41.5	18	510
BW-440	RO	100	0.68	1.00	-4.49	56.8	18	620
RE-BLR	RO	100	0.77	1.00	-20.90	46.8	15	480
NE-90	NF	200	2.17	0.98	-23.6	51.7	22	240
UTC-70	LPRO	100	0.99	1.00	-14.9	54.4	14	340
TS-80	NF	200	1.20	0.97	-14.00	48.0	4	500
Desal-HL	NF	300	2.00	0.97	-11.00	43.0	4	500
External								
ES-20	LPRO	100	1.69	1.00	-6.00	47.00	6	294
NF-70	NF	250	2.64	0.98	-25.00	29.8	18	1000
XLE-440	LPRO	150	0.92	0.98	-19.42	39.8	2	500
AK	LPRO	150	2.07	0.99	-20.00	50.00	23	800
ESPA	LPRO	200	1.44	0.99	-5.00	47.00	15	800
ESNA	LPRO	250	1.83	0.99	-9.90	63.2	22	800

a. 2000 mg/L MgSO$_4$, 25°C, recovery 15%, pressure 1034 kPa, pH 8.
b. Zeta potential (ZP) 10mM KCl, 10mM NaCl, pH 7 or 8.
PWP (pure water permeability); LPRO (low pressure reverse osmosis); CA (contact angle)

6.3 Physicochemical properties of organic compounds

Solute size descriptors considered for the models were molar volume (MV), molecular length, molecular width, molecular depth and equivalent molecular width. The authors selected molecular size descriptors based on previous studies (Kimura et al., 2004; Kiso et al., 2001b; Nghiem et al., 2004; Ozaki and Li, 2002; Schafer et al., 2003) that in some manner related rejection to size exclusion mechanisms between membranes and solutes. Truly, there are other topological descriptors of solute size that are more fundamental. However, a correlation between topological descriptors and molecular size exists and has been demonstrated by other studies. For instance, a previous study demonstrated the existence of a relationship between topological and topographical indices (Mihalic et al., 1992).

Another study of traditional topological indices, electronic and geometrical molecular descriptors used in QSAR research was conducted by Katritzky and Gordeeva (1993). They found that the best estimation of physicochemical properties was attained using classical topological indices such as the Randic index, Wiener index, and molecular connectivity indices; however, biological activity was better described by a combination of topological indices and geometrical descriptors. In addition, van der Waterbeemd et al. (1996) used molecular size descriptors, lipophilicity and hydrogen bonding capacity of solutes to unravel its contribution to membrane permeation estimation. In that sense, the validity of the selection of size descriptors as variables for membrane rejection prediction would be demonstrated.

Molecular Modeling Pro (ChemSW, Fairfield, CA) was used to compute molar volumes and molecular descriptors of size such as molecular length, width and depth, after optimization of the geometry of a molecule from the interaction of conformational analysis and energy minimization with the semi-empirical method MOPAC-PM3. Molecular weight (MW) was considered a fundamental descriptor; however, it is not an accurate descriptor of solute size because densities are different and geometrical configuration influences the size of a compound. The dipole moment, another fundamental descriptor, was calculated by Chem3D Ultra 7.0 (Cambridgesoft, Cambridge, UK).

The logarithm of the octanol-water partition coefficient (log K_{ow}) was used to describe solute hydrophobicity. Values of log K_{ow} were obtained from KOWWIN (EPA, USA). In fact, fundamental chemical parameters are used to predict log K_{ow}, and even neural network-based QSAR methods can produce better results than a group-contribution method as has been investigated by Yaffe et al. (2002). Nevertheless, the octanol-water partition coefficient was used as an empirical descriptor of hydrophobicity because various membrane studies have relied on the use of log K_{ow} as a hydrophobic descriptor (Kimura et al., 2003a; Kiso et al., 2001b; Nghiem et al., 2004; Schafer et al., 2003), and log K_{ow} values are easily accessible. Based on log K_{ow}, the compounds were classified as hydrophilic neutral (HL-neu) or hydrophobic neutral (HP-neu). Compounds with log $K_{ow} \geq 2$ were classified as hydrophobic, and those with log $K_{ow} < 2$ were classified as hydrophilic; more information about criteria used to classify compounds as hydrophobic or hydrophilic can be found elsewhere (Connell, 1990). The 50 neutral organic compounds in the database are shown in Table 6.1, together with calculated values of molecular weight, log K_{ow}, dipole moment, molecular length and equivalent width.

6.4   QSAR equation model and ANN models

The total number of variables (physicochemical descriptors, membrane characteristics, operating conditions) was reduced using a correlation matrix and factor analysis with principal component analysis (PCA). PCA is a process of variable reduction, whereby the redundancy present in the initial number of variables is reduced and variables that contain most of the variance are grouped into main components. A detailed PCA and MLR procedure has been presented in Chapter 5. During application of PCA there is a risk of selecting variables from the input space that may not be related to the output variable of MLR; the risk can be avoided by interpreting the relationtship of the selected variables with the output variable.

The database used in this chapter comprised 161 rejection cases for 50 compounds involving 15 initial variables. The membrane characteristics and operating conditions are presented in Table 6.3. The variables considered as solute descriptors were molecular weight (MW), log K_{ow}, dipole moment, molar volume, molecular length, molecular width, molecular depth and equivalent width. The variables describing membrane characteristics were molecular weight cut-off (MWCO), pure water permeability, magnesium sulphate salt rejection (SR), surface membrane charge (as zeta potential), and hydrophobicity (as contact angle); variables describing operating conditions were operating pressure and permeate flux. PCA was used to reduce the number of variables for the QSAR model. After PCA, a regression analysis was carried out using multiple linear regression (MLR). PCA and MLR were implemented using SPSS Statistics 16.0. Subsequently, the reduced number of variables defined by the QSAR equation allowed the development of an ANN that was used to predict rejections of neutral organic compounds by polyamide NF and RO membranes.

The intention of this chapter is the application of ANN as a computing modelling tool able to approximate a function; ANN can be used with that purpose (Rojas, 1996). It has been demonstrated that multi-layer feed-forward networks can be used as universal function approximators (Hornik et al., 1989). A summary of the theory and definitions related to ANN is presented in Appendix B, and applications on particular membrane studies has been presented in previous publications (Delgrange-Vincent et al., 2000 ; Cabassud et al., 2002 ; Zhao et al., 2005), and more detailed theories and definitions can be found in Haykin's book (1999). The ANN models were constructed after defining the QSAR model. The hypothesis is that the artificial neural network may improve the prediction ability of a QSAR equation (function). The importance of the QSAR step lies in the definition of the relevant variables. In this way, it was hypothesized that the performance of the QSAR model can be improved. ANN analyses were performed using the Neural Network Toolbox 5 of MATLAB 2007b and SPSS Neural Networks 16.0.

6.5   Results and discussion

6.5.1   Reduction of variables with PCA and QSAR model

After application of PCA, 15 variables were reduced to two components with six variables; component 1 was related to size and hydrophobicity with four variables (equivalent width, depth, length, log K_{ow}, all with positive loadings); component 2 was related to membrane characteristics with variables molecular weight cut-off (MWCO, positive loading) and salt rejection (SR, negative loading). Fig. 6.1a shows the loading plot of both components, revealing the clustering of variables that represents component 1 (size and hydrophobicity) and component 2 (membrane characteristics); SR and MWCO with negative and positive loadings, respectively, are shown in the figure in the lower and upper locations. Scores of the principal components are presented in Figure 6.1b, showing that two compound groups are distinguished in the graph. In general, hydrophobic neutral (HP-neu) and hydrophilic neutral (HL-neu) clustering is observed, however, not all cases cluster due to the influence of membrane characteristics on the components.

After obtaining the principal components, the following task was the implementation of MLR using the reduced set of variables and 97 random cases (60% of data) of rejection in order to model and test the equation. Two random sets of 32 rejection cases, S1 and S2 (20% of data each) were used as prediction sets. Although the selection of data for each group was made randomly, the rationale for data selection was that training, validation and prediction sets contained rejection cases over the entire range of rejections. The general QSAR linear equation for rejection ($R^2=0.81$) was as follows:

$$rejection = 183.920\,eqwidth + 31.830\,length - 0.549\,log\,K_{ow} + 883.294\,SR - 945.133 \quad (6.1)$$

A mechanistic interpretation of Eq. 6.1 is that rejection will increase with increasing length and equivalent width due to steric hindrance. Hydrophobicity, expressed as log K_{ow}, will decrease rejection due to adsorption and subsequent partitioning mechanisms. These relationships are in accordance with findings from the literature (Berg et al., 1997; Van der Bruggen et al., 2000; Kiso et al., 2000, 2001a, 2001b; Schafer et al., 2003;

Kimura et al. 2003b, 2004; Chen et al., 2004; Yoon et al., 2006; Agenson and Urase, 2007). The equation also shows that salt rejection of magnesium sulphate (SR) is a parameter incorporating steric/size hindrance and electrostatic repulsion effects related to the charge of the membrane and operating conditions. It is simpler to characterise a membrane in terms of salt rejection rather than MWCO. Moreover, MWCO alone is frequently unable to predict rejection (Kimura et al., 2004; Schafer et al., 2005). Therefore, a lump membrane parameter is needed for the prediction; thus, it is suggested that a specific parameter such as salt rejection may represent differences in charge, pore size and operating conditions between membranes, although the effect of hydrophobicity may not be captured. Nonetheless, the limitation of the QSAR model is that the constructed equation is valid for aromatic polyamide membranes and for boundary experimental conditions defined in Section 6.2.

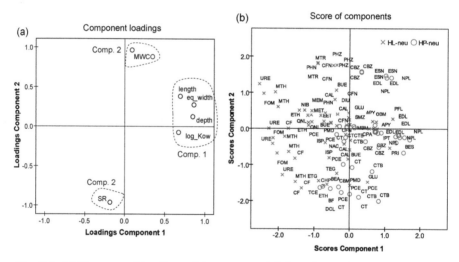

Fig. 6.1: (a) Component loadings plot, component 1 (length, equivalent width, log K_{ow}, depth), component 2 (MWCO, SR); (b) Scores of principal components

6.5.2   Artificial neural network models

The ANN models used in this study were multi-layer feed-forward back-propagation networks. The input layer contains the predictors and the hidden layer contains the number of neurons used; the output layer is the variable rejection. Table 6.4 shows the artificial neural network models used in this study. That table also shows inputs, number of neurons and graphical representations of each model. For all models, the hidden layer transfer function was a hyperbolic tangent sigmoid function. The output layer transfer function was a hyperbolic tangent sigmoid function for all models, except model N2 where a linear function was used. The output layer in all models was rejection. The training method used was Levenberg-Marquardt, except for model N2 where a scaled conjugate gradient training method was used. The performance of each model was evaluated in terms of a mean absolute percent error (MAPE), an average absolute error (AAE), a maximum absolute percent error (MaxAPE) and a standard deviation of error (STDE). They were determined using equations 6.2, 6.3, 6.4 and 6.5.

$$MAPE = \frac{1}{n}\sum_{i=1}^{n}\left|\frac{R_{pi} - R_{mi}}{R_{mi}}\right| \tag{6.2}$$

$$AAE = \left|R_{pi} - R_{mi}\right| \tag{6.3}$$

$$MaxAPE = \max\left|\frac{R_{pi} - R_{mi}}{R_{mi}}\right| \tag{6.4}$$

$$STDE = \sqrt{\frac{\sum_{i=1}^{n}\left(\frac{R_{pi} - R_{mi}}{100} - \frac{1}{n}\sum_{i=1}^{n}(\frac{R_{pi} - R_{mi}}{100})\right)^{2}}{n-1}} \tag{6.5}$$

where R_{pi} is predicted rejection, R_{mi} is measured rejection and n is the number of rejection cases.

The results of the performance of all models are shown in Table 6.5 (training, validation and prediction). The input database was randomly divided into three sets. The rationale for data selection was that training, validation and prediction sets contained rejection cases over the entire range of rejections. Sixty percent of the dataset was employed for training and 20% of the data was used for validation. Finally, the remaining 20% of data was used as an independent prediction dataset to test the network.

Table 6.4: Multi-layer feed-forward artificial neural networks

Name*	Input layer	N° neurons	Graphical representation	Software
N1	log K_{ow} length eqwidth SR	2		MATLAB 2007b
N2	log K_{ow} length eqwidth SR	2		SPSS 16
N3	log K_{ow} dipole length eqwidth SR	2		MATLAB 2007b
N4	log K_{ow} length eqwidth	2		MATLAB 2007b
N5	log K_{ow} length eqwidth SR	2		MATLAB 2007b
N6	log K_{ow} length eqwidth SR	4		MATLAB 2007b

* N1, N2, N3 and N4 (random dataset S1), N5 and N6 (random dataset S2)

Table 6.5: Performance of ANN and MLR models for training, validation and prediction

Model	Training				Validation				Prediction			
	MAPE	AAE	MaxAPE	STDE (%)	MAPE	AAE	MaxAPE	STDE (%)	MAPE	AAE	MaxAPE	STDE (%)
N1	0.154	6.113	2.897	7.3	0.681	5.36	8.074	6.9	0.328	4.56	8.074	5.4
N2	0.17	6.508	3.97	7.9	0.73	5.476	8.899	6.7	0.357	4.739	8.899	5.3
N3	0.147	6.091	2.73	7.2	0.813	6.182	10.36	8.0	0.436	6.005	10.36	7.1
N4	0.221	9.571	3.387	12.0	1.262	14.062	14.661	17.4	0.637	10.329	14.661	13.3
N5	0.27	5.712	4.498	7.1	0.165	6.032	1.596	7.4	0.139	5.059	1.596	6.2
N6	0.338	6.262	5.774	7.6	0.158	6.278	1.2	7.5	0.139	5.001	0.827	6.3
MLR*	0.254	8.751	6.799	10.8	1.318	9.789	12.772	11.7	0.545	7.346	12.529	8.5

MAPE mean absolute percent error
AAE average absolute error
MaxAPE maximum absolute percent error
STDE standard deviation of error
* training is model, validation is predictionS1, prediction is predictionS2

The prediction performance of the QSAR model (Eq. 6.1) is shown in Fig. 6.2 and Table 6.5. Hydrophobic neutral compounds showed rejection over 40%, and hydrophilic neutral compounds cover a broader range of rejection but with fewer high rejection cases. The main disadvantage of the MLR model is that it shows over- and under-prediction of rejection values in many cases (Fig. 6.2), although it yields acceptable correlation coefficients R^2 of 0.81 and 0.92, for the model dataset and prediction set S1, respectively. The standard deviation of error (STDE) was 10.8% for the model dataset and 11.7% for the prediction S1 set.

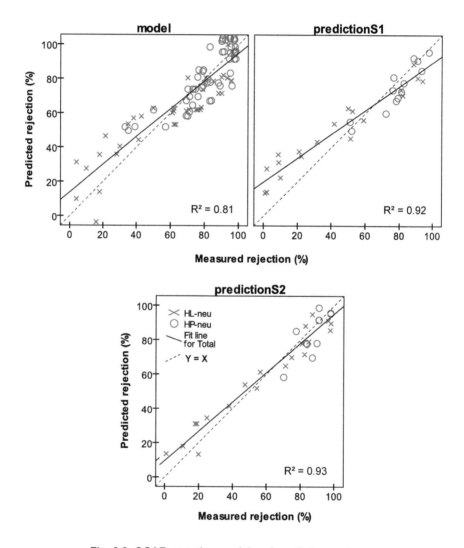

Fig. 6.2: QSAR equation model and predictions sets

The accuracy of prediction was improved by ANN models N1 and N2 (R^2 = 0.97) as can be observed in Fig. 6.3, Fig. 6.4 and Table 6.5, with an STDE of 5.4% and 5.3% for model N1 and N2, respectively. It is important to mention that different training methods and software were used to build models N1 and N2.

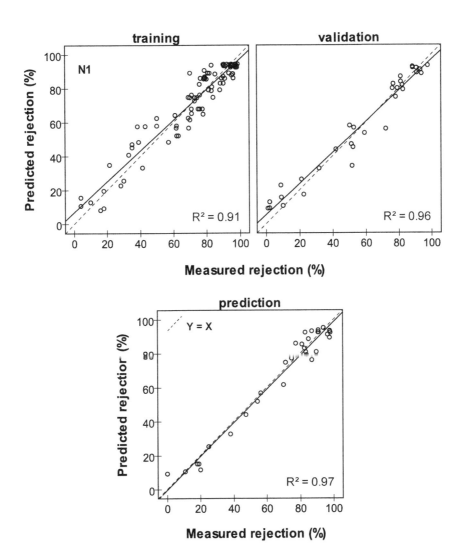

Fig. 6.3: Network model N1 with training, validation and prediction sets

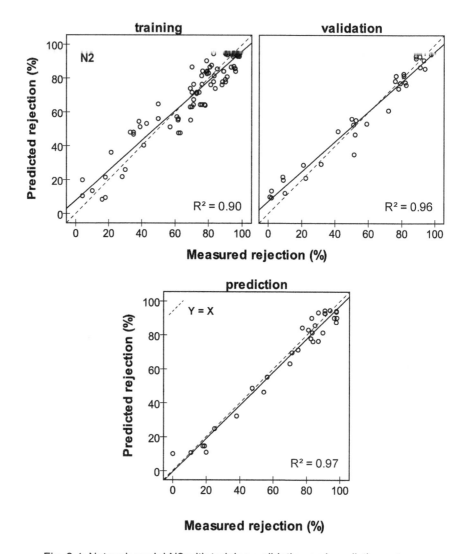

Fig. 6.4: Network model N2 with training, validation and prediction sets

In order to study the effect of inclusion of the dipole moment in the model, model N3 was implemented; it incorporated the dipole moment as an additional input. The result of model N3 is presented in Fig. 6.5, and it can be observed (Table 6.5) that the effect of dipole moment did not significantly improve the performance of network model N3 (STDE 7.1%) compared to models N1 (STDE 5.4%) and N2 (STDE 5.3%). A low or high dipole moment for a neutral compound is the result of the configuration of partial charges and their physical distribution to determine a vector sum of momentums. However, neutral compounds may only develop further interaction with the membrane in terms of size exclusion and partitioning that ultimately define rejection. Other membrane studies have identified dipole-dipole and hydrophobic interactions as influencing the passage of organic solutes through hydrophobic membranes, and interaction trends between a solute dipole moment and hydrophobicity with rejection have been presented (Kiso et al., 1992; Schafer et al., 2003; Kimura et al., 2003a; Chen et al., 2004; Bellona et al., 2004; Nghiem et al., 2005; Yoon et al., 2006).

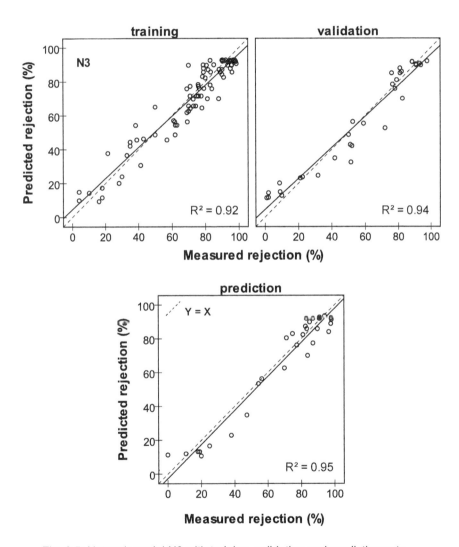

Fig. 6.5: Network model N3 with training, validation and prediction sets

An improvement in models N1 and N2 was found when compared to the QSAR model (STDE 8.5%). N1 and N2 included salt rejection as an input variable; however, a proof that salt rejection is an important input variable is required. In order to evaluate the importance of salt rejection as an input variable, a model without salt rejection was constructed (model N4). The performance results of network model N4 are shown in Fig. 6.6 and Table 6.5; as expected, the model without salt rejection presented the worst performance (model N4, STDE 13.3%) compared to the rest of the neural network models, and its performance was even inferior to the QSAR model (STDE 8.5%).

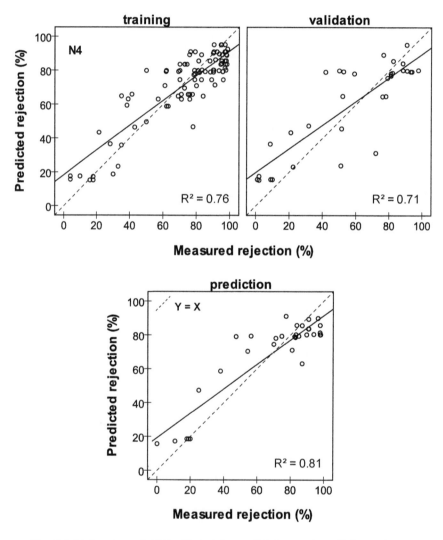

Fig. 6.6: Network model N4 with training, validation and prediction sets

Moreover, the importance of variables or their influence on network model N2 is shown in Fig. 6.7, which shows that SR was even more important than log K_{ow}. Their importance was calculated by SPSS 16; an explanation about the calculation procedure is beyond the scope of this thesis, however Papadokonstantakis et al. (2006) and Olden et al. (2004) elaborate more on the comparison of methods used to calculate the importance of variables on neural networks.

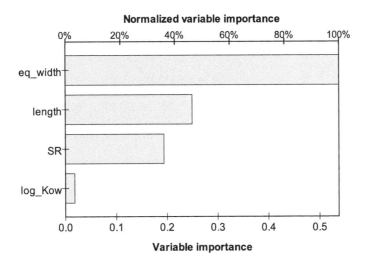

Fig. 6.7: Variable importance in network model N2

Finally, the effect of the number of neurons in the hidden layer was evaluated with network models N5 and N6, represented by Fig. 6.8 and Fig. 6.9, respectively. It was found that two neurons (N5) were sufficient in the hidden layer, and the effect of 4 neurons (N6) did not improve the predictions and performance (STDE 6.2% and 6.3% for N5 and N6, respectively). It is important to mention that models N1, N2, N3 and N4 used a random dataset S1, and models N5 and N6 used a different random dataset S2. Differences in random datasets for training, validation and prediction did not improve performance either. Even though it is not shown in Tables 6.4 and 6.5, it has to be mentioned that increasing the number of neurons to more than 4 resulted in over-fitting of the training group and increased standard deviation errors for the validation and prediction groups.

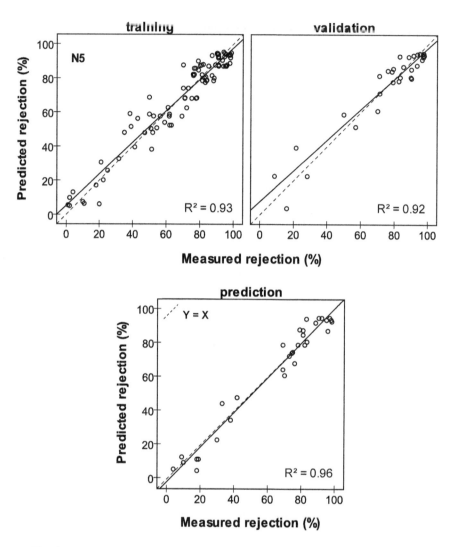

Fig. 6.8: Network model N5 with training, validation and prediction sets

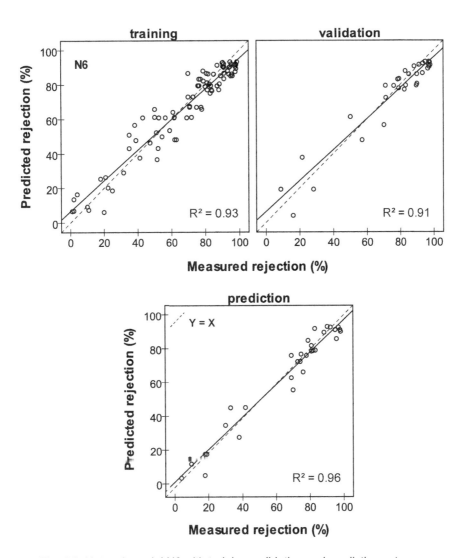

Fig. 6.9: Network model N6 with training, validation and prediction sets

6.6   Conclusions

o   Artificial neural networks may be an important tool for prediction of
    the rejection of neutral organic compounds by NF and RO
    membranes.

o   ANN models can be enhanced through the use of quantitative
    structure-activity relationships (QSAR) that may summarize
    interactions between membrane characteristics, filtration operating
    conditions and physicochemical properties of organic compounds.

o   It appeared that rejection of neutral organic compounds was mainly
    governed by size exclusion and hydrophobic interactions between
    the solute and membrane.

o   Magnesium sulphate salt rejection may be a possible lump parameter
    that defines the size exclusion capability of neutral organic
    compounds by NF or RO membranes; however, it may only be valid
    in combination with solute descriptors and for a range of boundary
    experimental conditions.

o   For the most promising ANN models, the independent predicted
    rejection values were compared to measured rejections and good
    correlations were found ($R^2=0.97$) and predicted rejections resulted
    in a standard deviation of error of ca. 5%.

Chapter 7

Implementation of NF as a robust barrier for organic contaminants during water reuse applications

Based on parts of:
- Is Nanofiltration a Robust Barrier for Organic Micropollutants, 5th IWA Specialized Membrane Technology Conference for Water and Wastewater Treatment, IWA Membrane Technology Conference & Exhibition, 1-3 September, 2009, Beijing, China.
- Nanofiltration as a robust barrier for organic micropollutants in water reuse, Submitted to Environmental Science & Technology (2009).

7.1   Introduction

Presently, it is not clear what the public health effects of mixtures of emerging organic contaminants in drinking water are, but it is evident that advanced drinking water treatment is needed to remove a majority of them to lower levels (1–100ng/L). The problem of removing organic contaminants increases when indirect water reuse projects are considered, especially in regions vulnerable to water scarcity (Rodriguez et al., 2009). Various studies have identified nanofiltration (NF) and reverse osmosis (RO) as proven technologies that remove most of the emerging organic contaminants (Van der Bruggen et al., 1999; Kiso et al., 2001b; Ozaki and Li, 2002; Schafer et al., 2003; Kimura et al., 2003b; Kimura et al., 2004; Nghiem et al., 2004). Although these studies were conducted at laboratory scale, they identified the main mechanisms of removal occurring through interactions between organic solutes and membranes: electrostatic repulsion, size/steric exclusion, hydrophobic adsorption and partitioning. Only a few studies have reported full-scale results of NF and RO for removal of emerging organic contaminants during water treatment for groundwater recharge or drinking water treatment (Radjenovic et al., 2008; Bellona et al., 2008; Schrotter et al., 2009). The trend in practical implementation of technology has favoured RO instead of NF, however, it is not clear why RO is preferred even though NF may show similar removal efficiencies or even better results in terms of operation and maintenance costs.

This chapter investigates why NF should be considered as a robust and more cost-effective barrier against emerging organic contaminants than RO. Secondly, we propose a modified concept of a multiple barrier approach considering aquifer recharge and recovery (a more economic treatment option) followed by NF, and alternative biodegradation and treatment options for organic compounds (1,4-dioxane, NDMA) that are difficult to remove with NF and RO.

7.2   Experimental

The experimental methodology for this chapter has been described in Sections 3.3 and 4.2. Nevertheless, for the sake of readability, the list of organic contaminants mentioned in this chapter (with their respective physicochemical properties) is presented in Table 7.1. The experimental methodology for laboratory-scale soil column studies, simulating aquifer recharge and recovery, was described by Maeng et al. (2008) with feeds of 1–2µg/L per compound.

Table 7.1: List of organic contaminants in feed water

Name	ID	MW (g/mol)	Log K_{ow}^a	Dipole (debye)[b]	MV^c (cm^3/mol)	Molec. length (nm)[c]	Equiv. width (nm)[c]	Effec. diam. (nm)[c]
Neutral								
Acetaminophen	ACT	151	0.46	4.55	120.90	1.14	0.53	0.79
Phenacetine	PHN	179	1.58	4.05	163.00	1.35	0.54	0.89
Caffeine	CFN	194	-0.07	3.71	133.30	0.98	0.70	0.77
Metronidazole	MTR	171	-0.02	6.30	117.80	0.93	0.66	0.75
Phenazone	PHZ	188	0.38	4.44	162.70	1.17	0.66	0.83
1,4-dioxane*	DIX	88	-0.27	0.00	89.10	0.71	0.59	0.57
Carbamazepine	CBM	236	2.45	3.66	186.50	1.20	0.73	0.89
17β-estradiol	E2	272	4.01	1.56	232.60	1.39	0.74	0.97
Estrone	E1	270	3.13	3.45	232.10	1.39	0.76	0.97
Bisphenol A	BPA	228	3.32	2.13	199.50	1.25	0.79	0.89
17α-ethynilestradiol	EE2	296	3.67	1.27	225.60	1.48	0.85	1.02
Ionic								
Sulphamethoxazole	SFM	253	0.89	7.34	173.10	1.33	0.64	0.89
Fenoprofen	FNP	242	-0.02	1.88	180.90	1.16	0.83	0.88
Ketoprofen	KTP	254	-0.52	3.42	187.90	1.16	0.83	0.87
Naproxen	NPN	230	3.18	2.55	192.20	1.37	0.76	0.93
Ibuprofen	IBF	206	3.97	4.95	200.30	1.39	0.64	0.93
Gemfibrozil	GFB	250	4.77	0.95	221.90	1.58	0.78	1.09

a. Experimental database: SRC PhysProp Database; b. Chem3D Ultra 7.0; c. Molecular Modeling Pro; * only present in NF-90 feed water

One of the purposes of this chapter is to compare the removal efficiencies of the membrane NF-90 determined at laboratory-scale with removals of NF-90 and RO membranes at pilot- and full-scale facilities. Our experimental work did not include NF and RO membranes at pilot- and full-scale facilities; therefore we used rejection results of three studies (Radjenovic et al., 2008; Bellona et al., 2008; Schrotter et al., 2009) that included RO membranes as well as the NF-90 membrane during pilot and full-scale tests. Experimental conditions and membrane characteristics of the bench, pilot and full-scale tests are shown in Table 7.2.

The organic contaminants with rejections by NF-90 and rejections by two RO membranes (BW30LE and ESPA2) are summarised in Table 7.3. Although the internal experiments were performed in a bench-scale setup, it is important to mention that concentration polarization during bench-scale experiments may be comparable to concentration polarization in membrane modules (elements with recoveries 10–16%), as demonstrated in Section 3.4.1.

Table 7.2: List of membranes and operating conditions used for comparison of NF and RO

Name	Test	MWCO (Da) SR (%)	Feed conc. (μg/L)	pH	Flux (L/m²-h)	Reco- very (%)	Pressure (kPa)	ZP* (mV)
NF-200*	BS	~300 75.0	5–18	7	13	8[†]	483	-28
NF-90*	BS	~200 90.0	5–18	7	13	8[†]	345	-32
NF-90* [a, b]	PS	~200 90.0	0.01–2.6	6–7	20[a]	10[b], 80[a]	n.a.	n.a.
NF-90* [c]	FS	~200 90.0	0.008–0.14	6	23	65	n.a.	n.a.
BW30LE* [b]	PS	<200 99.0	0.01–0.5	7	n.a.	10	n.a.	n.a.
BW30LE* [c]	FS	<200 99.0	0.008–0.14	6	23	73	n.a.	n.a.
ESPA2[□][a]	FS	<200 99.5	0.05–2.6	6–7	17	80	n.a.	n.a.

* Dow-Filmtec
□ Hydranautics
† Recovery for flat sheet membrane in BS setup. Other recoveries are total recoveries for PS and FS
BS (bench-scale)
PS (pilot-scale)
FS (full-scale)
SR (salt rejection of NaCl, 2,000mg/L, 480 kPa, 25°C and 15% recovery)
ZP (zeta potential at pH 7 and ionic strength 10mM KCl)
a. Bellona et al., 2008
b. Schrotter et al., 2009
c. Radjenovic et al, 2008

Table 7.3: Removal of compounds by NF and RO membranes

Compound	Abbr. (MW)	NF-90[a] rejections (%)	BW30LE[b] rejections (%)	ESPA2[c] rejections (%)
Neutral				
N-Nitrosodimethylamine	NDMA (74)	45	n.a.	35
1,4-Dioxane	DIX (88)	47	n.a.	n.a.
Acetaminophen	ACT (151)	67	72	n.a.
Metronidazole	MTR (171)	72	n.a.	n.a.
Phenacetine	PHN (179)	73	n.a.	83
Phenazone	PHZ (188)	93	n.a.	n.a.
Caffeine	CFN (194)	92	n.a.	90
Atrazine	ATZ (216)	97	96	n.a.
Bisphenol A	BPA (228)	91	n.a.	95
Carbamazepine	CBM (236)	95	99	95
Estrone	E1 (270)	93	n.a.	n.a.
17β-estradiol	E2 (272)	96	96	n.a.
17α-ethynylestradiol	EE2 (296)	93	95	n.a.
Ionic				
Salicylic acid	SAC (138)	98	n.a.	98
Ibuprofen	IBF (206)	97	n.a.	99
Naproxen	NPX (230)	97	n.a.	100
Fenoprofen	FNP (242)	95	n.a.	n.a.
Gemfibrozil	GFB (250)	97	n.a.	100
Sulphamethoxazole	SFM (253)	97	100	n.a.
Ketoprofen	KTP (254)	97	98	100
Diclofenac	DCF (296)	100	100	100

a. Average rejections of internal results and results from Bellona et al., 2008, Radjenovic et al, 2008, Schrotter et al., 2009
b. Average rejections of results from Radjenovic et al., 2008 and Schrotter et al., 2009
c. Rejections from Bellona et al., 2008

7.3   Results and discussion

7.3.1   Removal efficiencies by NF and RO membranes

According to the manufacturer of NF-90 (Dow-Filmtec), the membrane has a pore size of ~1nm. This membrane is intended to remove organics with molecular weights greater than 200g/mol while allowing varying amounts of mono (Na^+, Cl^-) and divalent (Ca^{2+}, Mg^{2+}) ion passage. Rejections of neutral and ionic compounds by NF-90 membranes are shown in Fig. 7.1. Rejections of neutral and ionic compounds by NF-200 and NF-90 are shown in Fig. 7.2; clearly it can be recognised that the low MWCO of NF-90 favoured (compared to NF-200) the increased rejection of sterically larger compounds with MW > 200g/mol.

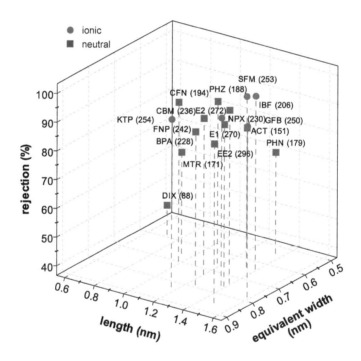

Fig. 7.1: Size and charge influence on removals (average of clean and fouled) of organic compounds by NF-90

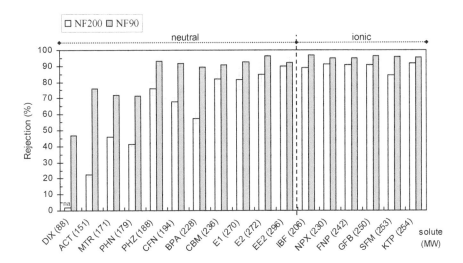

Fig. 7.2: Size and charge influence on removals of organic compounds by NF-200 and NF-90

Fouling of NF-90 membranes with NOM did not result in significant changes in the rejection of neutral and ionic compounds, with variations of - 2–4%. The removals presented in Figs. 7.1 and 7.2 are average rejections of tests with clean and NOM fouled membranes. Fig. 7.1 illustrates the relationship between rejection and size (length and equivalent width), MW (shown in parentheses) and classification (ionic and neutral) of the compounds. DIX (1,4-dioxane) exhibited the smallest length (0.71nm), effective diameter (0.57nm) and equivalent width (0.59nm), thus, size interactions between the membrane and solute resulted in low removal (47%) for 1.4-dioxane. It appeared that for the NF-90 membranes the effect of hydrophobicity of solutes and dipole moment had no impact on increase/decrease of rejection; however, NF-200 presented hydrophobic/partitioning interactions, especially with BPA, diminishing removals. PHZ is bigger in size (length and equivalent width) than CFN; thus, resulting in increased removal of the former by NF-200.

Rejection of negatively charged ionic compounds by NF-90 was higher (>95%) than NF-200 (>85%) due to effects of electrostatic interactions between the charge of the membrane surface and the charge of compounds and due to augmented removal by size exclusion mechanisms. The results confirmed that NF-90 removed most of the emerging contaminants by size exclusion due to a combined effect of length > 1nm and equivalent width > 0.7nm, with rejections over 90%. A calculation of the pore size of NF-90 using rejection results of compounds with rejections (89–91%) revealed an average pore size of 0.9nm, slightly less than that given by the manufacturer (~1nm). The pore size was estimated as the aritmetic mean of the effective diameters with rejections in the range 89–91%. The average pore size for NF-200 was estimated as 1.1nm. Fig. 7.1 illustrates that removal of neutral compounds was influenced by their size. Small compounds (151 < MW < 200) exhibited acceptable rejection: ~75%. Although NF is not effective for removing 1,4-dioxane (~45% removal), water reuse applications should verify that the contaminant is actually present in the feed water. In contrast, all ionic compounds showed high rejections (>95%). The overall rejection (considering 1,4-dioxane) for clean and fouled membranes was 81% and 83%, respectively, for all compounds. The overall rejection without 1,4-dioxane for clean and fouled membranes was 84% and 86%, respectively. Individual rejections and the overall (summation) rejection were calculated as follows

$$R = 1 - \frac{C_p}{C_f} \tag{7.1}$$

$$R_{all} = 1 - \frac{\sum C_p}{\sum C_f} \tag{7.2}$$

where C_p is the concentration of each solute in the permeate and C_f is the concentration of each solute in the feed. R is rejection per solute and R_{all} is the overall rejection of compounds present in the cocktail.

Laboratory results of this study showed that the overall rejection of NF-90 was about 82% and 95% for neutral and ionic compounds, respectively (Fig. 7.2), and they compared to pilot and full-scale results with overall rejections of about 83% and 99% for neutral and ionic compounds, respectively (data from references in Table 7.2), thus it may be possible to predict the rejection of ionic and neutral compounds in full-scale plants based on lab-scale experimental results, although more trials may be required on different water sources and under different operating conditions.

In order to demonstrate that NF is an effective barrier for most of the emerging contaminants present in different source waters, a comparison of the average rejection of NF-90 and two RO membranes (BW30LE and ESPA2) is illustrated in Figs. 7.3 and 7.4 employing average rejection data from laboratory-, pilot- and full-scale tests. Fig. 7.3 shows that 7 neutral compounds with molecular weights of less than 200g/mol were tested with RO membranes at pilot- or full-scale installations. For instance, ACT was included during a full-scale test (Radjenovic et al., 2008) and exhibited rejection of ~72%, whereas the rejection with NF-90 was ~67%. Similarly, the difference in rejection for PHN was only 10% between NF-90 and ESPA2.

The investigation by Bellona et al. (2008) tested the removal of NDMA; experiments from a pilot test showed 45% removal by NF-90, and a full-scale test showed NDMA removal of 35% with ESPA2. Compounds with a MW > 200 were removed by NF-90, BW30LE and ESPA2 with any rejections over ~90%. In Fig. 7.4 the removal of ionic compounds is presented and it is clear that NF and RO membranes are quite effective even for low molecular weight ionic compounds (e.g. salicylic acid, SAC). Hence, laboratory-, pilot- and full-scale tests show that NF-90 is a membrane able to remove emerging organic contaminants with almost the same effectiveness as BW30LE and ESPA2, both RO membranes with MWCO < 200 and SR (NaCl) of ~99%.

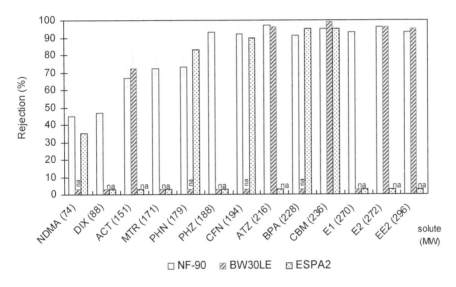

Fig. 7.3: Comparison of rejections of neutral compounds by NF and RO

An additional demonstration of the effectiveness of NF in removing micropollutants is its ability to remove pesticides. Rejection of 7 pesticides was reported in Appendix D and E (atrazine, bentazon, chlorotoluron, isoproturon, lindane, metoxuron and simazine). Taking into consideration individual feed concentrations of 0.5µg/L and percentage rejections by NF-90 and TS-80 (NF, Trisep, MWCO ~200), individual pesticide concentrations of the permeate were less than or equal to 0.1µg/L, hence they comply with the European Union directive of drinking water quality (Directive 98/83/EC). Moreover, the total pesticide concentration in the permeate resulted in 0.35µg/L, and also complies with the directive that allows a total pesticide concentration of less than 0.5µg/L. The same approach may be implemented for the regulation of representative PhACs and EDCs. Calculations with individual feed concentrations of 0.5µg/L for PhACs and EDCs shown in Table 7.1, and average rejections by NF-90, resulted in individual permeate concentrations of less than 0.15μg/L and 0.06μg/L for PhACs and EDCs, respectively; and a total permeate concentration of 0.5μg/L and 0.16μg/L for PhACs and EDCs, respectively.

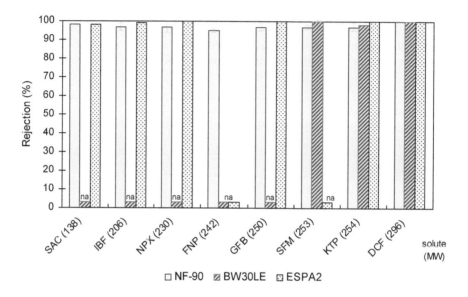

Fig. 7.4: Comparison of rejections of ionic compounds by NF and RO

7.3.2 Cost analysis and water reuse facilities

The economical comparison of using NF instead of RO membranes was also made. For this purpose, a water reuse treatment plant for reclamation of secondary treated effluent was practically designed using existing information of completed projects (Schippers et al., 2004) and non-commercial software (ROSA, Dow-Filmtec). The software predicts the pump-membrane energy consumption of a nanofiltration (NF-90) and reverse osmosis (BW30LE) membrane of the same manufacturer (Dow-Filmtec). The feed water characteristics and the configuration for both trains of the plant were assumed to be the same. The plant was designed for production of 100 m³/h of permeate; the plant comprises 1 skid with a two-stage configuration; 12:6 pressure vessels in the first pass, and 8:6 pressure vessels in the second stage, with each pressure vessel accommodating six 8" elements. No post-treatment of produced water was considered in the cost analysis assuming that the water will not be *directly* reused.

The results comparing capital expenditure (CAPEX) and operation and maintenance (O&M) for NF and RO are shown in Table 7.4. The analysis of treatment costs revealed that the total cost of treatment per cubic metre of water with NF was lower ($0.55/m³) than RO ($0.57/m³), which means $0.02/m³ savings.

Table 7.4: Analysis of investment and treatment costs for NF and RO

	NF90-400			BW30LE-440		
	CAPEX	O&M		CAPEX	O&M	
Item	Invest.	Annual	Total	Invest.	Annual	Total
	$*	$*	$/m³	$*	$*	$/m³
Capital expenditure						
Buildings & site development	329			329		
Membrane gen. & pre-treatment	988			988		
Membranes RO/NF elements	84			72		
Electric	439			439		
Pumping system & reservoir	219			219		
Raw & finished water piping	104			104		
Concentrate disposal	110			110		
Contingency	417			417		
Sub total	2,690			2,678		
Amortization CAPEX $/m³ †	0.29		0.29	0.29		0.29
O&M costs						
Energy, NF/RO membranes ‡		19			27	
Energy, others		55			55	
Chemicals		29			33	
Membrane replacement		8			7	
Concentrate disposal		30			35	
Maintenance & labor		86			86	
Sub total		227			243	
O&M $/m³		0.26	0.26		0.28	0.28
Total $/m³			0.55			0.57

* ×1000
† 20 years, interest rate of 7%
‡ ROSA 7 (Dow, Filmtec), 0.24kWh (NF-90), 0.33kWh (BW30LE), $0.09/kWh.

A key issue is that most designers put water-quality considerations in first place not being aware that NF has also similar quality capabilities. Taking into account that the investment cost of NF is $12k more expensive than RO, those costs can be recovered with savings in O&M in almost a year. An NF-90 membrane (8"×40") element is more expensive ($100) than a BW30LE membrane element, and this trend is also valid for other NF and RO membranes. After some membrane marketing analysis and communication with manufacturers, this difference can be explained as being due to two reasons: i) different NF membrane fabrication, ii) high demand for RO membranes resulting in lower prices. However, new projects may decide that the initial high investment in NF membrane costs can be recovered in a few years, with lower expense in O&M in the long term. A hypothetical decrease in the price of NF membrane per element to that of RO element will decrease the CAPEX but will not have a big effect on the amortized cost of water. According to the manufacturer a NF90-400 element has a thicker feed spacer (34mil) than a BW30LE-440 element (28mil). A hypothetical increase in NF membrane area by 40ft^2 (3.8m^2)/element with 28mil feed spacer will reduce the number of elements and, combined with the effect of a lower price (equal to RO), will save an additional 0.01$/m^3.

Several full-scale applications of water reuse for groundwater recharge rely on the use of RO membranes. In Belgium, the Torreele facility treats the effluent of the Wulpen wastewater treatment plant (WWTP) combining the use of ultrafiltration (ZeeWeed 500c) and RO (BW30LE) membranes for the treatment of 370m^3/h of wastewater effluent (Van Houtte and Verbauwhede, 2007). Other examples of water reuse at a larger capacity are the NEWater project in Singapore, and the Orange County Water Replenishment Groundwater (WRG). The NEWater plants will treat 23,921m^3/h at full operation for 2010 (Rohe et al., 2009), the treatment processes involve microfiltration, ultrafiltration, RO, and UV disinfection. Formerly known as Water Factory 21 with a capacity of (~789m^3/h for RO), the project has developed into the WRG project able to treat 11,039m^3/h. In Orange County, a microfiltration treatment is used for pre-treatment of a secondary wastewater effluent, that later undergoes RO treatment followed by an advanced oxidation process (AOP) of ultraviolet light with hydrogen peroxide (Patel, 2009).

Figs. 7.5 and 7.6 illustrate a process schematic of the treatment for Torreele and Orange County, respectively. It is important to mention that for Orange County, UV in combination with hydrogen peroxide is intended to remove only NDMA and 1,4-dioxane present in contaminated feed waters. All emerging organic contaminants presently being monitored in WRG (9 endocrine disruptors and 10 pharmaceuticals) are removed to below detection limits in the process; this treatment scheme is known as the multiple barrier approach. Nonetheless, the cost of treatment is expensive; after some calculations (20 years, interest rate of 7%) the cost of treatment in WRG is ~$0.78/m^3 (Patel, 2009). The cost of treatment for Torreele has been reported as $0.60/m^3 (Van Houtte and Verbauwhede, 2007).

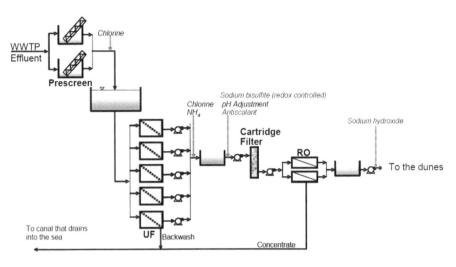

Fig. 7.5: Schematic process at Torreele treatment plant

(Van Houte and Verbauwhede, 2007)

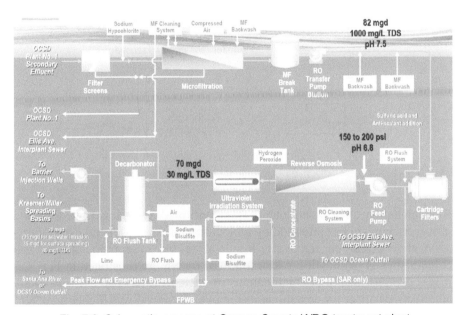

Fig. 7.6: Schematic process at Orange County WRG treatment plant

(Patel, 2009)

7.3.3   Modified multiple barrier approach for water reuse

A modified concept of the multiple barrier approach may result in a more cost-effective solution with NF. In order to make NF part of the modified concept of a multiple barrier approach, a new schematic treatment process is suggested in Fig. 7.6, whereby aquifer recharge and recovery (ARR) before NF would increase the rate of removal of micropollutants. Additional control of contamination at the source for 1,4-dioxane and NDMA will remove or at least reduce the presence of contaminants during further water treatment.

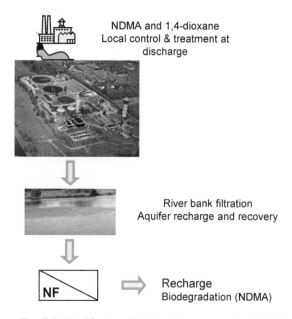

Fig. 7.6: Modified multiple barrier approach with NF

Biodegradation of micropollutants was confirmed as removal phenomenon in soil column studies after adding a biocide (sodium azide) and determining a negative effect on removals. Removal results of compounds for soil column studies (simulating ARR) followed by NF are illustrated in Fig. 7.7. It was demonstrated that combinations of ARR with NF-200 and NF-90 are an effective barrier for contaminants removal. Removal was higher than 90% for ANN-NF90 and higher than 80% for ANN-NF200. NF provides synergy to ARR, both acting as an effective barrier. Carbamazepine (CBM) a compound not removed by ARR (0%) was well removed by NF-200 and NF-90.

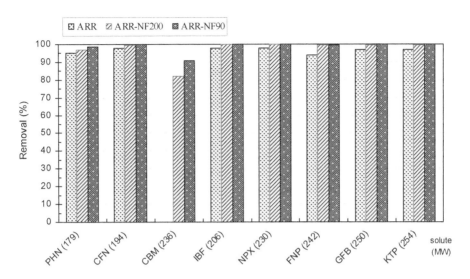

Fig. 7.7: Removal of organic contaminants with ARR-NF (modified multiple barrier approach)

N-nitrosodimethylamine (NDMA) is an organic compound of concern in water reuse applications; it has been reported that NDMA showed low rejections (35–45%) with NF and RO membranes (Plumlee et al., 2008). NDMA is classified as a B2 carcinogen, reasonably anticipated to be a human carcinogen and is present in the draft candidate list of drinking water contaminants (EPA, 2008a; EPA, 2008b). NDMA has been reported to be formed as a disinfection by-product (DBP) during the disinfection of secondary-treated wastewater effluent with chloramines; NDMA is formed from the reaction of monochloramine and organic nitrogen-containing precursors (Mitch et al., 2003; Mitch and Sedlak, 2004). A recent publication that reports field observations of groundwater exposed to water recharge of recycled water containing NDMA revealed that the compound was able to biodegrade (80%) in 626 days during a localized discharge-extraction event (Zhou et al., 2009). Evidence for biodegradation was supported by temporal changes in the observed magnitude and extent of NDMA in groundwater, and temporal changes in the estimated regional mass in groundwater consistent with removal of NDMA mass and groundwater transport modelling calibration processes. More importantly, an experimental laboratory study using soils under aerobic and anaerobic conditions demonstrated biodegradation of NDMA (Bradley et al., 2005). Their results indicated that NDMA disappeared after 132 days of experiments under aerobic conditions, with *Pseudomonas putida* and *Pseudomonas fluorescens* being responsible for the degradation under those conditions; for anaerobic conditions, 67–78% of NDMA was biodegraded, while for a sterile soil control no loss of NDMA was observed.

During NF and RO treatment for water reuse, another contaminant of interest is 1,4-dioxane, originating from the use of solvents, stabilizing agents, surfactants (detergents, cosmetics, foods), household aerosol products, water proofing aerosol products, paint and varnish strippers and by manufacturing of polyester (Zenker et al., 2003). 1,4-dioxane is classified as a Class B2 (a probable) human carcinogen (EPA), and its presence has been confirmed in many water environments (Abe, 1999). The compound is also special because it is highly soluble and quite resistant to biodegradation, characteristics related to its structure; a cyclic organic compound containing two symmetrically opposed ether linkages (Zenker et al., 2003).

A previous investigation demonstrated that 1,4-dioxane was degraded to below detection (~5µg/L) when it was incubated aerobically at 35°C with tetrahydrofuran (TFC) acting as a potential co-substrate (Zenker et al., 2000). The co-metabolic treatment process of degradation of 1,4-dioxane using THF as a co-substrate was also investigated using a laboratory-scale rotating biological filter that showed sustained and feasible biodegradation of 1,4-dioxane at low concentrations (1–30 mg/L); additionally, a laboratory-scale trickling filter with THF maintained a 95–98% removal rate for over one year with a feed solution containing 0.2–2.5mg/L of 1,4-dioxane (Zenker et al., 2002; Zenker et al., 2004). Another treatment alternative for 1,4-dioxane is the use of phytoremediation; 80% of 1,4-dioxane was removed by hybrid poplars, and enhanced bioaugmentation of hybrid poplars with the micro-organism Amycolata CB1990 resulted in mineralization of 1,4-dioxane (Aitchison et al., 2000; Kelley et al., 2001). Although implementation of this technique in water reuse schemes lessens the recovery of water for indirect reuse, still it is an economic alternative for 1,4-dioxane control.

7.4   Conclusions

o   NF is an effective barrier for micropollutants, its removal performance compares to RO. Only a few small organic contaminants are not well removed by NF.

o   NF is less expensive than RO due to savings in energy consumption; this allows reduced costs of O&M. Additional savings can be obtained when the prices of NF membrane elements are less than RO elements.

o   Aquifer recharge and recovery followed by nanofiltration is an effective combination that will remove micropollutants first by biodegradation and later by the membrane through mechanisms of size exclusion and charge repulsion.

o   Biodegradation of NDMA can be achieved during groundwater recharge.

o   Source control of 1,4-dioxane contamination, and implementation in WWTP of alternative treatment processes to remove this compound, will help to reduce the presence of this contaminant during further water treatment.

Chapter 8

Recommendations

8.1   Recommendations for future research

- o It was demonstrated that most of the emerging organic contaminants can be well removed by tight NF membranes (~0.9nm pore size or MWCO ~200). However, small organic compounds (chloroform, 1,4-dioxane, NDMA) were not removed even by tight NF membranes. Therefore, when those compounds are present it is important to define strategies for their removal. More experimental work (pilot-scale) is needed to determine the real magnitude of removal under practical conditions of wastewater treatment plant effluents and the treatment of surface waters.

- o It is necessary to standardise protocols for the measurement of membrane salt rejection and MWCO. Those two parameters are important, and the success of models that rely on them will depend on standard protocols.

- o Experimental work (laboratory-, pilot- and full-scale) that considers additional treatment that involves alternative and cost-effective biodegradation processes will help to determine the degree to which it is possible to remove persistent organic contaminants that pass through NF membranes and even pass through RO membranes.

- o Another approach that can be experimented in the laboratory is the use of beds of activated carbon after NF experiments using less tight NF membranes. This approach still relies on the good capacity of some NF membranes (NF-200) to remove the major percentage of organic contaminants; moreover, the NF reduces the organic carbon loading on the activated carbon.

- o It would be beneficial to confirm the modelling results of this thesis developing QSAR-ANN models for existing NF and RO full-scale facilities. Monitoring programs for removal of emerging organic contaminants of importance should be carried out considering current practices of membrane treatment plants and real feed water matrices.

- o It is necessary to demonstrate that separate QSAR models will perform better for defined groups of membranes: polyamide "loose" NF membranes, polyamide "tight" NF membranes and low pressure RO membranes.

- o Additional experimental demonstrations (bench-, pilot- and full-scale) of sequential water treatment by aquifer recharge and recovery followed by nanofiltration (loose and tight membranes) is needed in order to support that a modified multiple barrier approach for removal of micropollutants is more cost-effective than other options (RO and AOP).

References

Abdi, H., 2003. Partial Least Squares (PLS) Regression in Encyclopedia of Social Sciences Research Methods. Lewis-Beck, M., Bryman, A. and Futing, T. (eds), Sage, Thousand Oaks, CA.

Abe, A., 1999. Distribution of 1,4-dioxane in relation to possible sources in the water environment. The Science of the Total Environment 227(1), 41-47.

Adams, C., Wang, Y., Loftin, K. and Meyer, M., 2002. Removal of antibiotics from surface and distilled water in conventional water treatment processes. Journal of Environmental Engineering-Asce 128(3), 253-260.

Agenson, K.O., Oh, J.-I. and Urase, T., 2003. Retention of a wide variety of organic pollutants by different nanofiltration/reverse osmosis membranes: controlling parameters of process. Journal of Membrane Science 225(1-2), 91-103.

Agenson, K.O. and Urase, T., 2007. Change in membrane performance due to organic fouling in nanofiltration (NF)/reverse osmosis (RO) applications. Separation and Purification Technology 55(2), 147-156.

Aitchison, E.W., Kelley, S.L., Alvarez, P.J.J. and Schnoor, J.L., 2000. Phytoremediation of 1,4-Dioxane by hybrid poplar trees. Water Environment Research 72(3), 313-321.

Baker, R.W., 2004. Membrane technology and applications, John Wiley & Sons Ltd., Southern Gate, England.

Bartels, C., Wilf, M., Casey, W. and Campbell, J., 2008. New generation of low fouling nanofiltration membranes. Desalination 221(1-3), 158-167.

Bellona, C. and Drewes, J.E., 2005. The role of membrane surface charge and solute physico-chemical properties in the rejection of organic acids by NF membranes. Journal of Membrane Science 249(1-2), 227-234.

Bellona, C. and Drewes, J.E., 2007. Viability of a low-pressure nanofilter in treating recycled water for water reuse applications: A pilot-scale study. Water Research 41(17), 3948-3958.

Bellona, C., Drewes, J.E., Oelker, G., Luna, J., Filteau, G. and Amy, G., 2008. Comparing nanofiltration and reverse osmosis for drinking water augmentation. Journal American Water Works Association 100(9), 102-116.

Bellona, C., Drewes, J.E., Xu, P. and Amy, G., 2004. Factors affecting the rejection of organic solutes during NF/RO treatment - a literature review. Water Research 38(12), 2795-2809.

158

Ben-David, A., Bason, S., Jopp, J., Oren, Y. and Freger, V., 2006. Partitioning of organic solutes between water and polyamide layer of RO and NF membranes: Correlation to rejection. Journal of Membrane Science 281(1-2), 480-490.

Benotti, M.J., Trenholm, R.A., Vanderford, B.J., Holady, J.C., Stanford, B.D. and Snyder, S.A., 2008. Pharmaceuticals and Endocrine Disrupting Compounds in U.S. Drinking Water. Environmental Science & Technology 43(3), 597-603.

Berg, P., Hagmeyer, G. and Gimbel, R., 1997. Removal of pesticides and other micropollutants by nanofiltration. Desalination 113(2-3), 205-208.

Bicknell, R.J., Herbison, A.E. and Sumpter, J.P., 1995. Estrogenic Activity of an Environmentally Persistent Alkylphenol in the Reproductive-Tract but Not the Brain of Rodents. Journal of Steroid Biochemistry and Molecular Biology 54(1-2), 7-9.

Bowen, W.R. and Mohammad, A.W., 1998. Characterization and Prediction of Nanofiltration Membrane Performance--A General Assessment. Chemical Engineering Research and Design 76(8), 885-893.

Boyd, G.R., Reemtsma, H., Grimm, D.A. and Mitra, S., 2003. Pharmaceuticals and personal care products (PPCPs) in surface and treated waters of Louisiana, USA and Ontario, Canada. Science of The Total Environment 311(1-3), 135-149.

Bradley, P.M., Carr, S.A., Baird, R.B. and Chapelle, F.H., 2005. Biodegradation of N-nitrosodimethylamine in soil from a water reclamation facility. Bioremediation Journal 9(2), 115 - 120.

Braghetta, A., DiGiano, F.A. and Ball, W.P., 1997. Nanofiltration of natural organic matter: pH and ionic strength effects. Journal of Environmental Engineering 123(7), 628-641.

Brian, P.L.T., 1966. Mass transport in reverse osmosis in Desalination by reverse osmosis. Merten, Ulrich (ed), MIT Press, Cambridge.

Cabassud, M., Delgrange-Vincent, N., Cabassud, C., Durand-Bourlier, L. and Lainé, J.M., 2002. Neural networks: a tool to improve UF plant productivity. Desalination 145(1-3), 223-231.

Campbell, P., Srinivasan, R., Knoell, T., Phipps, D., Ishida, K., Safarik, J., Cormack, T. and Ridgway, H., 1999. Quantitative structure-activity relationship (QSAR) analysis of surfactants influencing attachment of a Mycobacterium sp. to cellulose acetate and aromatic polyamide reverse osmosis membranes. Biotechnology and Bioengineering 64(5), 527-544.

Cardew, P.T. and Le, M.S., 1998. Membrane Processes: A Technology Guide, The Royal Society of Chemistry, Cambridge.

Castiglioni, S., Bagnati, R., Fanelli, R., Pomati, F., Calamari, D. and Zuccato, E., 2006. Removal of pharmaceuticals in sewage treatment plants in Italy. Environmental Science & Technology 40(1), 357-363.

Chen, S.-S., Taylor, J.S., Mulford, L.A. and Norris, C.D., 2004. Influences of molecular weight, molecular size, flux, and recovery for aromatic pesticide removal by nanofiltration membranes. Desalination 160(2), 103-111.

Childress, A.E. and Elimelech, M., 2000. Relating nanofiltration membrane performance to membrane charge (electrokinetic) characteristics. Environmental Science & Technology 34(17), 3710-3716.

Chiou, C.T., 2002. Partition and adsorption of organic contaminants in environmental systems by Cary T. Chiou, Wiley-Interscience, Hoboken.

Cho, J., Amy, G. and Pellegrino, J., 2000. Membrane filtration of natural organic matter: factors and mechanisms affecting rejection and flux decline with charged ultrafiltration (UF) membrane. Journal of Membrane Science 164(1-2), 89-110.

Choi, K.J., Kim, S.G., Kim, C.W. and Kim, S.H., 2005. Effects of activated carbon types and service life on removal of endocrine disrupting chemicals: amitrol, nonylphenol, and bisphenol-A. Chemosphere 58(11), 1535-1545.

Clara, M., Strenn, B., Gans, O., Martinez, E., Kreuzinger, N. and Kroiss, H., 2005. Removal of selected pharmaceuticals, fragrances and endocrine disrupting compounds in a membrane bioreactor and conventional wastewater treatment plants. Water Research 39(19), 4797-4807.

Cleuvers, M., 2004. Mixture toxicity of the anti-inflammatory drugs diclofenac, ibuprofen, naproxen, and acetylsalicylic acid. Ecotoxicology and Environmental Safety 59(3), 309-315.

Comerton, A.M., Andrews, R.C., Bagley, D.M. and Hao, C.Y., 2008. The rejection of endocrine disrupting and pharmaceutically active compounds by NF and RO membranes as a function of compound a water matrix properties. Journal of Membrane Science 313(1-2), 323-335.

Connell, D.W., 1990. Bioaccumulation of Xenobiotic Compounds, CRC Press, Boca Raton, FL.

Cornelissen, E.R., Verdouw, J., Gijsbertsen-Abrahamse, A.J. and Hofman, J.A.M.H., 2005. A nanofiltration retention model for trace contaminants in drinking water sources. Desalination 178(1-3), 179-192.

Dearden, J.C. and Ghafourian, T., 1999. Hydrogen Bonding Parameters for QSAR: Comparison of Indicator Variables, Hydrogen Bond Counts, Molecular Orbital and Other Parameters. Journal of Chemical Information and Computer Sciences 39(2), 231-235.

Delgrange, N., Cabassud, C., Cabassud, M., Durand-Bourlier, L. and Laine, J.M., 1998. Modelling of ultrafiltration fouling by neural network. Desalination 118(1-3), 213-227.

Delgrange-Vincent, N., Cabassud, C., Cabassud, M., Durand-Bourlier, L. and Laîné, J.M., 2000. Neural networks for long term prediction of fouling and backwash efficiency in ultrafiltration for drinking water production. Desalination 131(1-3), 353-362.

Demuth, H., Beale, M. and Hagan, M., 2007. Neural Network Toolbox User's Guide, The MathWorks, Inc., Natick, MA.

Drewes, J.E., Heberer, T. and Reddersen, K., 2002. Fate of pharmaceuticals during indirect potable reuse. Water Science and Technology 46(3), 73-80.

Drewes, J.E., Amy, G., Kim, T.U., Xu, P., Bellona, C., Oedekoven, M. and Macalady, D., 2006. Rejection of Wastewater-Derived Micropollutants in High-Pressure Membrane Applications Leading to Indirect Potable Reuse Effect of Membranes and Micropollutant Properties. WateReuse.

Elimelech, M., Chen, W.H. and Waypa, J.J., 1994. Measuring the Zeta (Electrokinetic) Potential of Reverse-Osmosis Membranes by a Streaming Potential Analyzer. Desalination 95(3), 269-286.

Ellis, J.B., 2006. Pharmaceutical and personal care products (PPCPs) in urban receiving waters. Environmental Pollution 144(1), 184-189.

EPA, 1998. Endocrine Disruptor Screening and Testing Advisory Committee (EDSTAC) Final Report. Environmental Protection Agency.

EPA, 2008a. Drinking Water Contaminant Candidate List 3—Draft, EPA.

EPA, 2008b. Emerging Contaminant - N-Nitroso-dimethylamine (NDMA). EPA (ed).

Eriksson, L., Jaworska, J., Worth, A.P., Cronin, M.T., McDowell, R.M. and Gramatica, P., 2003. Methods for reliability and uncertainty assessment and for applicability evaluations of classification- and regression-based QSARs. Environmental Health Perspective 111(10), 1361-1375.

Escher, B.I., Bramaz, N., Eggen, R.I. and Richter, M., 2005. In vitro assessment of modes of toxic action of pharmaceuticals in aquatic life. Environmental Science & Technology 39(9), 3090-3100.

European Parliament, 1999. Community Strategy for Endocrine Disruptors, a range of substances suspected of interfering with the hormone systems of humans and wildlife, European Parliament, COM (1999) 706.

European Parliament, 2000. Water Framework Directive, European Parliament, Directive 2000/60/EC.

European Parliament, 2001. Establishing the list of priority substances in the field of water policy, European Parliament, Decision No 2455/2001/EC.

European Parliament, 2006. Proposal for a Directive of the European Parliament and of the Council on environmental quality standards in the field of water policy and amending, European Parliament, Directive 2000/60/EC. COM(2006) 398.

Everitt, B.S. and Dunn, G., 2001. Applied Multivariate Data Analysis, Arnold, London.

FDA, 1998. Guidance for industry environmental assessment of human drug and biologics applications. Food and Drug Administration.

Fent, K., Weston, A.A. and Caminada, D., 2006. Ecotoxicology of human pharmaceuticals. Aquatic Toxicology 76(2), 122-159.

Fujikawa, M., Nakao, K., Shimizu, R. and Akamatsu, M., 2007. QSAR study on permeability of hydrophobic compounds with artificial membranes. Bioorganic & Medicinal Chemistry 15(11), 3756-3767.

Ghafourian, T. and Dearden, J.C., 2004. The use of molecular electrostatic potentials as hydrogen-bonding-donor parameters for QSAR studies. Il Farmaco 59(6), 473-479.

Gramatica, P., 2007. Principles of QSAR models validation: internal and external. QSAR & Combinatorial Science 26(5), 694-701.

Guillette, L.J., Gross, T.S., Masson, G.R., Matter, J.M., Percival, H.F. and Woodward, A.R., 1994. Developmental Abnormalities of the Gonad and Abnormal Sex-Hormone Concentrations in Juvenile Alligators from Contaminated and Control Lakes in Florida. Environmental Health Perspectives 102(8), 680-688.

Hagan, M.T., Demuth, H.B. and Beale, M.H., 2002. Neural network design, Martin Hagan.

Haykin, S., 1999. Neural networks a comprehensive foundation, Prentice-Hall, Upper Saddle River.

Heberer, T., 2002. Occurrence, fate, and removal of pharmaceutical residues in the aquatic environment: a review of recent research data. Toxicology Letters 131(1-2), 5-17.

Henderson, R.K., Baker, A., Parsons, S.A. and Jefferson, B., 2008, Characterisation of algogenic organic matter extracted from cyanobacteria, green algae and diatoms. Water Research 42(13), 3435-3445.

Her, N., Amy, G. and Jarusutthirak, C., 2000. Seasonal variations of nanofiltration (NF) foulants: identification and control. Desalination 132(1-3), 143-160.

Hirsch, R., Ternes, T., Haberer, K. and Kratz, K.L., 1999. Occurrence of antibiotics in the aquatic environment. Science of The Total Environment 225(1-2), 109-118.

Hoek, E.M.V., Kim, A.S. and Elimelech, M., 2002. Influence of Crossflow Membrane Filter Geometry and Shear Rate on Colloidal Fouling in Reverse Osmosis and Nanofiltration Separations. Environmental Engineering Science 19(6), 357-372.

Hoek, E.M. and Elimelech, M., 2003. Cake-enhanced concentration polarization: a new fouling mechanism for salt-rejecting membranes. Environmental Science & Technology 37(24), 5581-5588.

Hoek, E.M.V., Bhattacharjee, S. and Elimelech, M., 2003. Effect of membrane surface roughness on colloid-membrane DLVO interactions. Langmuir 19(11), 4836-4847.

Hornik, K., Stinchcombe, M. and White, H., 1989. Multilayer feedforward networks are universal approximators. Neural Networks 2(5), 359-366.

Howe, K.J., Ishida, K.P. and Clark, M.M., 2002. Use of ATR/FTIR spectrometry to study fouling of microfiltration membranes by natural waters. Desalination 147(1-3), 251-255.

Hua, W.Y., Bennett, E.R. and Letcher, R.J., 2006. Ozone treatment and the depletion of detectable pharmaceuticals and atrazine herbicide in drinking water sourced from the upper Detroit River, Ontario, Canada. Water Research 40(12), 2259-2266.

Jackson, J. and Sutton, R., 2008. Sources of endocrine-disrupting chemicals in urban wastewater, Oakland, CA. Science of The Total Environment 405(1-3), 153-160.

Jarusutthirak, C., Amy, G. and Croué, J.-P., 2002. Fouling characteristics of wastewater effluent organic matter (EfOM) isolates on NF and UF membranes. Desalination 145(1-3), 247-255.

Johnson, R.A. and Wichern, D.W., 2007. Applied Multivariate Statistical Analysis, Pearson Prentice Hall, Upper Saddle River, NJ.

Jolliffe, I.T., 2002. Principal Component Analysis, Springer-Verlag, New York.

Jones, O.A., Lester, J.N. and Voulvoulis, N., 2005. Pharmaceuticals: a threat to drinking water? Trends Biotechnology 23(4), 163-167.

Jørgensen, B. and Goegebeur, Y., 2006. Multivariate Data Analysis and Chemometrics, Department of Statistics, University of Southern Denmark, Odense.

Katritzky, A.R. and Gordeeva, E.V., 1993. Traditional topological indexes vs electronic, geometrical, and combined molecular descriptors in QSAR/QSPR research. Journal of Chemical Information and Computer Sciences 33(6), 835-857.

Kelley, S.L., Aitchison, E.W., Deshpande, M., Schnoor, J.L. and Alvarez, P.J.J., 2001. Biodegradation of 1,4-dioxane in planted and unplanted soil: effect of bioaugmentation with amycolata sp. CB1190. Water Research 35(16), 3791-3800.

Khan, S.J., Wintgens, T., Sherman, P., Zaricky, J. and Schafer, A.I., 2004. Removal of hormones and pharmaceuticals in the advanced water recycling demonstration plant in Queensland, Australia. Water Science and Technology 50(5), 15-22.

Kim, H.-C. and Dempsey, B.A., 2008. Effects of wastewater effluent organic materials on fouling in ultrafiltration. Water Research 42(13), 3379-3384.

Kim, T.U., Amy, G. and Drewes, J.E., 2005. Rejection of trace organic compounds by high-pressure membranes. Water Science and Technology 51(6-7), 335-344.

Kim, T.U., Drewes, J.E., Summers, R.S. and Amy, G.L., 2007. Solute transport model for trace organic neutral and charged compounds through nanofiltration and reverse osmosis membranes. Water Research 41(17), 3977-3988.

Kimura, K., Amy, G., Drewes, J. and Watanabe, Y., 2003a. Adsorption of hydrophobic compounds onto NF/RO membranes: an artifact leading to overestimation of rejection. Journal of Membrane Science 221(1-2), 89-101.

Kimura, K., Amy, G., Drewes, J.E., Heberer, T., Kim, T.U. and Watanabe, Y., 2003b. Rejection of organic micropollutants (disinfection by-products, endocrine disrupting compounds, and pharmaceutically active compounds) by NF/RO membranes. Journal of Membrane Science 227(1-2), 113-121.

Kimura, K., Toshima, S., Amy, G. and Watanabe, Y., 2004. Rejection of neutral endocrine disrupting compounds (EDCs) and pharmaceutical active compounds (PhACs) by RO membranes. Journal of Membrane Science 245(1-2), 71-78.

Kiso, Y., Kitao, T., Jinno, K. and Miyagi, M., 1992. The Effects of Molecular Width on Permeation of Organic Solute through Cellulose-Acetate Reverse-Osmosis Membranes. Journal of Membrane Science 74(1-2), 95-103.

Kiso, Y., Nishimura, Y., Kitao, T. and Nishimura, K., 2000. Rejection properties of non-phenylic pesticides with nanofiltration membranes. Journal of Membrane Science 171(2), 229-237.

Kiso, Y., Kon, T., Kitao, T. and Nishimura, K., 2001a. Rejection properties of alkyl phthalates with nanofiltration membranes. Journal of Membrane Science 182(1-2), 205-214.

Kiso, Y., Sugiura, Y., Kitao, T. and Nishimura, K., 2001b. Effects of hydrophobicity and molecular size on rejection of aromatic pesticides with nanofiltration membranes. Journal of Membrane Science 192(1-2), 1-10.

Kolpin, D.W., Furlong, E.T., Meyer, M.T., Thurman, E.M., Zaugg, S.D., Barber, L.B. and Buxton, H.T., 2002. Pharmaceuticals, hormones, and other organic wastewater contaminants in US streams, 1999-2000: A national reconnaissance. Environmental Science & Technology 36(6), 1202-1211.

Kröse, B. and Van der Smagt, P., 1996. An Introduction to Neural Networks, University of Amsterdam, Faculty of Mathematics & Computer Science, Amsterdam.

Landau, S. and Everitt, B., 2004. A Handbook of Statistical Analysis using SPSS, Chapman & Hall/CRC, Boca Raton.

Lee, N., Amy, G., Croue, J.P. and Buisson, H., 2004. Identification and understanding of fouling in low-pressure membrane (MF/UF) filtration by natural organic matter (NOM). Water Research 38(20), 4511-4523.

Lee, S., Park, G., Amy, G., Hong, S.K., Moon, S.H., Lee, D.H. and Cho, J., 2002. Determination of membrane pore size distribution using the fractional rejection of nonionic and charged macromolecules. Journal of Membrane Science 201(1-2), 191-201.

Lee, S., Ang, W.S. and Elimelech, M., 2006. Fouling of reverse osmosis membranes by hydrophilic organic matter: implications for water reuse. Desalination 187(1-3), 313-321.

Li, Q. and Elimelech, M., 2004. Organic fouling and chemical cleaning of nanofiltration membranes: measurements and mechanisms. Environmental Science & Technology 38(17), 4683-4693.

Li, Q.L., Xu, Z.H. and Pinnau, I., 2007. Fouling of reverse osmosis membranes by biopolymers in wastewater secondary effluent: Role of membrane surface properties and initial permeate flux. Journal of Membrane Science 290(1-2), 173-181.

Libotean, D., Giralt, J., Rallo, R., Cohen, Y., Giralt, F., Ridgway, H.F., Rodriguez, G. and Phipps, D., 2008. Organic compounds passage through RO membranes. Journal of Membrane Science 313(1-2), 23-43.

Ma, S., Song, L., Ong, S.L. and Ng, W.J., 2004. A 2-D streamline upwind Petrov/Galerkin finite element model for concentration polarization in spiral wound reverse osmosis modules. Journal of Membrane Science 244(1-2), 129-139.

Maeng, S.K., Sharma, S.K., Amy, G. and Magic-Knezev, A., 2008. Fate of effluent organic matter (EfOM) and natural organic matter (NOM) through riverbank filtration. Water Science & Technology 57(12), 1999-2007.

Majewska-Nowak, K., Kabsch-Korbutowicz, M. and Dodz, M., 2002. Effects of natural organic matter on atrazine rejection by pressure driven membrane processes. Desalination 145(1-3), 281-286.

Makdissy, G., Peldszus, S., McPhail, R. and Huck, P.M., 2007. Towards a mechanistic understanding of the impact of fouling on the removal of EDCs/PCPs by nanofiltration membranes. AWWA Water Quality Technology Conference.

Matsui, Y., Knappe, D.R.U., Iwaki, K. and Ohira, H., 2002a. Pesticide adsorption by granular activated carbon adsorbers. 2. Effects of pesticide and natural organic matter characteristics on pesticide breakthrough curves. Environmental Science & Technology 36(15), 3432-3438.

Matsui, Y., Knappe, D.R.U. and Takagi, R., 2002b. Pesticide adsorption by granular activated carbon adsorbers. 1. Effect of natural organic matter preloading on removal rates and model simplification. Environmental Science & Technology 36(15), 3426-3431.

Mihalic, Z., Nikolic, S. and Trinajstic, N., 1992. Comparative study of molecular descriptors derived from the distance matrix. Journal of Chemical Information and Computer Sciences 32(1), 28-37.

Mitch, W.A., Sharp, J.O., Trussell, R.R., Valentine, R.L., Alvarez-Cohen, L. and Sedlak, D.L., 2003. N-Nitrosodimethylamine (NDMA) as a Drinking Water Contaminant: A Review. Environmental Engineering Science 20(5), 389-404.

Mitch, W.A. and Sedlak, D.L., 2004. Characterization and Fate of N-Nitrosodimethylamine Precursors in Municipal Wastewater Treatment Plants. Environmental Science & Technology 38(5), 1445-1454.

Mulder, M., 1996. Basic principles of membrane technology, Kluwer Academic Publishers, Dordrecht.

Ng, H.Y. and Elimelech, M., 2004. Influence of colloidal fouling on rejection of trace organic contaminants by reverse osmosis. Journal of Membrane Science 244(1-2), 215-226.

Nghiem, L.D., Schafer, A.I. and Waite, T.D., 2002. Adsorption of estrone on nanofiltration and reverse osmosis membranes in water and wastewater treatment. Water Science and Technology 46(4-5), 265-272.

Nghiem, L.D., Schafer, A.I. and Elimelech, M., 2004. Removal of natural hormones by nanofiltration membranes: Measurement, modeling, and mechanisms. Environmental Science & Technology 38(6), 1888-1896.

Nghiem, L.D., Schafer, A.I. and Elimelech, M., 2005. Pharmaceutical retention mechanisms by nanofiltration membranes. Environmental Science & Technology 39(19), 7698-7705.

Nghiem, L.D., Schafer, A.I. and Elimelech, M., 2006. Role of electrostatic interactions in the retention of pharmaceutically active contaminants by a loose nanofiltration membrane. Journal of Membrane Science 286(1-2), 52-59.

Nghiem, L., Espendiller, C. and Braun, G., 2008a. Influence of organic and colloidal fouling on the removal of sulphamethoxazole by nanofiltration membranes. World Water Conference and Exhibition 2008, 7-12 September.

Nghiem, L.D., Vogel, D. and Khan, S., 2008b. Characterising humic acid fouling of nanofiltration membranes using bisphenol A as a molecular indicator. Water Research 42(15), 4049-4058.

Olden, J.D., Joy, M.K. and Death, R.G., 2004. An accurate comparison of methods for quantifying variable importance in artificial neural networks using simulated data. Ecological Modelling 178(3-4), 389-397.

Ollers, S., Singer, H.P., Fassler, P. and Muller, S.R., 2001. Simultaneous quantification of neutral and acidic pharmaceuticals and pesticides at the low-ng/1 level in surface and waste water. Journal of Chromatography A 911(2), 225-234.

Orchin, M. and University of Cincinnatl. Dept. of Chemistry, 2005. The vocabulary and concepts of organic chemistry Online resource, Wiley-Interscience, Hoboken.

Ozaki, H. and Li, H., 2002. Rejection of organic compounds by ultra-low pressure reverse osmosis membrane. Water Research 36(1), 123-130.

Papadokonstantakis, S., Lygeros, A. and Jacobsson, S.P., 2006. Comparison of recent methods for inference of variable influence in neural networks. Neural Networks 19(4), 500-513.

Patel, M., 2009. Membrane technology takes reclamation to the limit. AWWA Membrane Technology Conference & Exhibition, 15-18 March.

Peng, X., Yu, Y., Tang, C., Tan, J., Huang, Q. and Wang, Z., 2008. Occurrence of steroid estrogens, endocrine-disrupting phenols, and acid pharmaceutical residues in urban riverine water of the Pearl River Delta, South China. Science of The Total Environment 397(1-3), 158-166.

Plumlee, M.H., López-Mesas, M., Heidlberger, A., Ishida, K.P. and Reinhard, M., 2008. N-nitrosodimethylamine (NDMA) removal by reverse osmosis and UV treatment and analysis via LC-MS/MS. Water Research 42(1-2), 347-355.

Pomati, F., Castiglioni, S., Zuccato, E., Fanelli, R., Vigetti, D., Rossetti, C. and Calamari, D., 2006. Effects of a complex mixture of therapeutic drugs at environmental levels on human embryonic cells. Environmental Science & Technology 40(7), 2442-2447.

Porter, M.C., 1972. Concentration polarization with membrane ultrafiltration. Industrial & Engineering Chemistry Product Research and Development 11(3), 234-248.

Radjenovic, J., Petrovic, M., Ventura, F. and Barceló, D., 2008. Rejection of pharmaceuticals in nanofiltration and reverse osmosis membrane drinking water treatment. Water Research 42(14), 3601-3610.

Ren, S., Das, A. and Lien, E.J., 1996. QSAR analysis of membrane permeability to organic compounds. Journal Drug Targeting 4(2), 103-107.

Rodriguez, C., Van Buynder, P., Lugg, R., Blair, P., Devine, B., Cook, A. and Weinstein, P., 2009. Indirect Potable Reuse: A Sustainable Water Supply Alternative. International Journal of Environmental Research and Public Health 6(3), 1174-1203.

Rohe, D.L., Koh, S.T. and Cairns, J., 2009. Case study of Singapore's largest operating advanced water reclamation plant: the Ulu Pandan NEWater plant. AWWA Membrane Technology Conference & Exhibition, 15-18 March.

Rojas, R., 1996. Neural Networks A Systematic Introduction, Springer, Berlin.

Roudman, A.R. and DiGiano, F.A., 2000. Surface energy of experimental and commercial nanofiltration membranes: Effects of wetting and natural organic matter fouling. Journal of Membrane Science 175(1), 61-73.

Sacher, F., Lange, F.T., Brauch, H.J. and Blankenhorn, I., 2001. Pharmaceuticals in groundwaters analytical methods and results of a monitoring program in Baden-Wurttemberg, Germany. Journal of Chromatography A 938(1-2), 199-210.

Sacher, F., Ehmann, M., Gabriel, S., Graf, C. and Brauch, H.J., 2008. Pharmaceutical residues in the river Rhine - results of a one-decade monitoring programme. Journal of Environmental Monitoring 10(5), 664-670.

Sangster, J., 1997. Octanol-Water Partition Coefficients: Fundamentals and Physical Chemistry, John Wiley & Sons, Chichester.

Santos, J.L.C., de Beukelaar, P., Vankelecom, I.F.J., Velizarov, S. and Crespo, J.G., 2006. Effect of solute geometry and orientation on the rejection of uncharged compounds by nanofiltration. Separation and Purification Technology 50(1), 122-131.

Sawyer, C.N., McCarty, P.L. and Parkin, G.F., 2003. Chemistry for environmental engineering and science, McGraw-Hill, Boston.

Schaep, J., Van der Bruggen, B., Vandecasteele, C. and Wilms, D., 1998. Influence of ion size and charge in nanofiltration. Separation and Purification Technology 14(1-3), 155-162.

Schaep, J. and Vandecasteele, C., 2001. Evaluating the charge of nanofiltration membranes. Journal of Membrane Science 188(1), 129-136.

Schafer, A.I., Nghiem, L.D. and Waite, T.D., 2003. Removal of the natural hormone estrone from aqueous solutions using nanofiltration and reverse osmosis. Environmental Science & Technology 37(1), 182-188.

Schafer, A.I., Fane, A.G. and Waite, T.D., 2005. Nanofiltration principles and applications, Elsevier Advanced Technology, Oxford.

Schippers, J.C., Kruithof, J.C., Nederlof, M.M., Hofman, J.A.M.H. and Taylor, J.S., 2004. Integrated membrane systems, AWWA Research Foundation, Denver, CO.

Schlett, C. and Pfeifer, B., 1996. Bestimmung von Steroidhormonen in Trink- und Oberflächenwässer = Determination of steroidal in drinking and surface water samples. Vom Wasser 87, 327-333.

Schrotter, J.-C., Breda, C., Vince, F., De Roubin, M.-R., Roche, P. and Bréant, P., 2009. Development of advanced treatment trains and decision support tools to meet future drinking water challenges in TECHNEAU: Safe drinking water from source to tap, state of the art & pespectives. Van den Hoven, Theo and Kazner, Christian (eds), IWA Publishing, London, UK.

168

Schwarzenbach, R.P., Gschwend, P.M. and Imboden, D.M., 2003. Environmental organic chemistry. John Wiley & Sons, New Jersey.

Shim, Y., Lee, H.J., Lee, S., Moon, S.H. and Cho, J., 2002. Effects of natural organic matter and ionic species on membrane surface charge. Environmental Science & Technology 36(17), 3864-3871.

Snyder, S.A., Westerhoff, P., Yoon, Y. and Sedlak, D.L., 2003. Pharmaceuticals, personal care products, and endocrine disruptors in water: Implications for the water industry. Environmental Engineering Science 20(5), 449-469.

Snyder, S.A., Adham, S., Redding, A.M., Cannon, F.S., DeCarolis, J., Oppenheimer, J., Wert, E.C. and Yoon, Y., 2007. Role of membranes and activated carbon in the removal of endocrine disruptors and pharmaceuticals. Desalination 202(1-3), 156-181.

Snyder, S.A., Lei, H. and Wert, E.C., 2008. Removal of Endocrine Disruptors and Pharmaceuticals during Water Treatment in Fate of Pharmaceuticals in the Environment and in Water Treatment Systems. Aga, D.S. (ed), CRC Press Taylor and Francis Group, Boca Raton.

Stasinakis, A.S., Gatidou, G., Mamais, D., Thomaidis, N.S. and Lekkas, T.D., 2008. Occurrence and fate of endocrine disrupters in Greek sewage treatment plants. Water Research 42(6-7), 1796-1804.

Stumpf, M., Ternes, T.A., Wilken, R.D., Rodrigues, S.V. and Baumann, W., 1999. Polar drug residues in sewage and natural waters in the state of Rio de Janeiro, Brazil. Science of The Total Environment 225(1-2), 135-141.

Sumpter, J.P. and Johnson, A.C., 2005. Lessons from endocrine disruption and their application to other issues concerning trace organics in the aquatic environment. Environmental Science & Technology 39(12), 4321-4332.

Taylor, J.S., Chen, S.S., Mulford, L.A. and Norris, C.D., 2000. Flat Sheet, Bench and Pilot Testing for Pesticide Removal Using Reverse Osmosis. AWWA Research Foundation - KIWA.

Ternes, T.A., 1998. Occurrence of drugs in German sewage treatment plants and rivers. Water Research 32(11), 3245-3260.

Ternes, T.A., Joss, A. and Siegrist, H., 2004. Scrutinizing pharmaceuticals and personal care products in wastewater treatment. Environmental Science & Technology 38(20), 392a-399a.

Ternes, T.A., Meisenheimer, M., McDowell, D., Sacher, F., Brauch, H.J., Gulde, B.H., Preuss, G., Wilme, U. and Seibert, N.Z., 2002. Removal of pharmaceuticals during drinking water treatment. Environmental Science & Technology 36(17), 3855-3863.

Thorpe, K.L., Maack, G., Benstead, R. and Tyler, C.R., 2009. Estrogenic Wastewater Treatment Works Effluents Reduce Egg Production in Fish. Environmental Science & Technology 43(8), 2976-2982.

Van de Ven, W.J.C., Sant, K.v.t., Pünt, I.G.M., Zwijnenburg, A., Kemperman, A.J.B., Van der Meer, W.G.J. and Wessling, M., 2008.

Hollow fiber dead-end ultrafiltration: Influence of ionic environment on filtration of alginates. Journal of Membrane Science 308(1-2), 218-229.

Van de Waterbeemd, H., Camenisch, G., Folkers, G. and Raevsky, O.A., 1996. Estimation of Caco-2 Cell Permeability using Calculated Molecular Descriptors. Quantitative Structure-Activity Relationships 15(6), 480-490.

Van der Bruggen, B., Schaep, J., Maes, W., Wilms, D. and Vandecasteele, C., 1998. Nanofiltration as a treatment method for the removal of pesticides from ground waters. Desalination 117(1-3), 139-147.

Van der Bruggen, B., Schaep, J., Wilms, D. and Vandecasteele, C., 1999. Influence of molecular size, polarity and charge on the retention of organic molecules by nanofiltration. Journal of Membrane Science 156(1), 29-41.

Van der Bruggen, B., Schaep, J., Wilms, D. and Vandecasteele, C., 2000. A Comparison of Models to Describe the Maximal Retention of Organic Molecules in Nanofiltration. Separation Science and Technology 35(2), 169 - 182.

Van der Bruggen, B. and Vandecasteele, C., 2002. Modelling of the retention of uncharged molecules with nanofiltration. Water Research 36(5), 1360-1368.

Van Houtte, E. and Verbauwhede, J., 2007. Torreele's water re-use facility enabled sustainable groundwater management in de Flemish dunes (Belgium). 6th IWA Specialist Conference Wastewater Reclamation and Reuse for Sustainability, 9-12 October.

Verliefde, A., Cornelissen, E., Amy, G., Van der Bruggen, B. and van Dijk, H., 2007. Priority organic micropollutants in water sources in Flanders and the Netherlands and assessment of removal possibilities with nanofiltration. Environmental Pollution 146(1), 281-289.

Verliefde, A.R.D., Cornelissen, E.R., Heijman, S.G.J., Verberk, J.Q.J.C., Amy, G.L., Van der Bruggen, B. and van Dijk, J.C., 2008. The role of electrostatic interactions on the rejection of organic solutes in aqueous solutions with nanofiltration. Journal of Membrane Science 322(1), 52-66.

Vieno, N., Tuhkanen, T. and Kronberg, L., 2006. Removal of pharmaceuticals in drinking water treatment: Effect of chemical coagulation. Environmental Technology 27(2), 183-192.

Vosges, M., Braguer, J.C. and Combarnous, Y., 2008. Long-term exposure of male rats to low-dose ethynylestradiol (EE2) in drinking water: effects on ponderal growth and on litter size of their progeny. Reproductive Toxicology 25(2), 161-168.

Wade, L.G., 2003. Organic chemistry by L.G. Wade, Jr, Prentice-Hall, Upper Saddle River.

170

Webb, S., Ternes, T., Gibert, M. and Olejniczak, K., 2003. Indirect human exposure to pharmaceuticals via drinking water. Toxicology Letters 142(3), 157-167.

Weinberg, H.S., Pereira, V.J. and Ye, Z., 2008. Drugs in Drinking Water Treatment Options in Fate of pharmaceuticals in the environment and in water treatment systems. Aga, D.S. (ed), CRC Press Taylor and Francis Group, Boca Raton.

Westerhoff, P., Yoon, Y., Snyder, S. and Wert, E., 2005. Fate of endocrine-disruptor, pharmaceutical, and personal care product chemicals during simulated drinking water treatment processes. Environmental Science & Technology 39(17), 6649-6663.

Wijmans, J.G, Athayde, A.L, Daniels, R., Ly, J.H., Kamaruddin, H.D. and Pinnau, I., 1996. The role of boundary layers in the removal of volatile organic compounds from water by pervaporation. Journal of Membrane Science 109(1), 135-146.

Wilson, B.A., Smith, V.H., Denoyelles, F. and Larive, C.K., 2003. Effects of three pharmaceutical and personal care products on natural freshwater algal assemblages. Environmental Science & Technology 37(9), 1713-1719.

Xu, P., Drewes, J.E., Bellona, C., Amy, G., Kim, T.U., Adam, M. and Heberer, T., 2005. Rejection of emerging organic micropollutants in nanofiltration-reverse osmosis membrane applications. Water Environment Research 77(1), 40-48.

Xu, P., Drewes, J.E., Kim, T.U., Bellona, C. and Amy, G., 2006. Effect of membrane fouling on transport of organic contaminants in NF/RO membrane applications. Journal of Membrane Science 279(1-2), 165-175.

Xu, Y.Z. and Lebrun, R.E., 1999. Investigation of the solute separation by charged nanofiltration membrane: effect of pH, ionic strength and solute type. Journal of Membrane Science 158(1-2), 93-104.

Yaffe, D., Cohen, Y., Espinosa, G., Arenas, A. and Giralt, F., 2002. Fuzzy ARTMAP and back-propagation neural networks based quantitative structure-property relationships (QSPRs) for octanol-water partition coefficient of organic compounds. Journal of Chemical Information and Computer Sciences 42(2), 162-183.

Yangali Quintanilla, V., 2005. Colloidal and non-colloidal NOM fouling of ultrafiltration membranes: analyses of membrane fouling and cleaning. MSc thesis SE 05-013, UNESCO-IHE, Institute for Water Education, Delft, the Netherlands.

Yoon, Y. and Lueptow, R.M., 2005. Removal of organic contaminants by RO and NF membranes. J Memb Sci 261(1-2), 76-86.

Yoon, Y., Westerhoff, P., Snyder, S.A. and Wert, E.C., 2006. Nanofiltration and ultrafiltration of endocrine disrupting compounds, pharmaceuticals and personal care products. Journal of Membrane Science 270(1-2), 88-100.

Zenker, M.J., Borden, R.C. and Barlaz, M.A., 2000. Mineralization of 1,4-dioxane in the presence of a structural analog. Biodegradation 11(4), 239-246.

Zenker, M.J., Borden, R.C. and Barlaz, M.A., 2002. Modeling Cometabolism of Cyclic Ethers. Environmental Engineering Science 19(4), 215-228.

Zenker, M.J., Borden, R.C. and Barlaz, M.A., 2003. Occurrence and Treatment of 1,4-Dioxane in Aqueous Environments. Environmental Engineering Science 20(5), 423-432.

Zenker, M.J., Borden, R.C. and Barlaz, M.A., 2004. Biodegradation of 1,4-Dioxane Using Trickling Filter. Journal of Environmental Engineering 130(9), 926-931.

Zhao, Y., Taylor, J.S. and Chellam, S., 2005. Predicting RO/NF water quality by modified solution diffusion model and artificial neural networks. Journal of Membrane Science 263(1-2), 38-46.

Zhou, Q., McCraven, S., Garcia, J., Gasca, M., Johnson, T.A. and Motzer, W.E., 2009. Field evidence of biodegradation of N-Nitrosodimethylamine (NDMA) in groundwater with incidental and active recycled water recharge. Water Research 43(3), 793-805.

Zorita, S., Mårtensson, L. and Mathiasson, L., 2009. Occurrence and removal of pharmaceuticals in a municipal sewage treatment system in the south of Sweden. Science of the Total Environment 407(8), 2760-2770.

List of symbols and abbreviations

Symbol	Unit	Description
K_{ow}		Octanol-water partition coefficient
$log\ K_{ow}$		Logarith of octanol-water partition coefficient
K_a		Acid dissociation constant
pK_a		Logarithmic form of acid dissociation constant
R^2		Regression coefficient squared
k	cm/s	Mass transfer coefficient
J	cm/s	Water permeation flux
J/k		Peclet number
c_m	µg/L	Concentration of solute at the membrane surface
c_b	µg/L	Concentration of solute in the feed solution (bulk)
c_p	µg/L	Concentration of solute in the permeate
δ	cm	Thickness of boundary layer
E		Enrichment factor
E_0		Enrichment factor at the membrane
J_0	cm/s	Initial pure water permeation flux
Q_p	cm³/s	Pure water permeate flow rate
A_m	m²	Membrane surface area
U	cm/s	Average cross-flow velocity
Q_c	cm³/s	Concentrate flow rate
D	cm²/s	Diffusion coefficient of the solute in water
d_h	cm	Equivalent hydraulic diameter
L	cm	Channel length
η	10^{-2} g/cm-s	Viscosity of water
MV	cm³/mol	Molar volume
C_p	µg/L	Permeate concentration
C_f	µg/L	Feed concentration
C_{p0}	µg/L	Initial permeate concentration
C_{f0}	µg/L	Initial feed concentration
DOC_m	mg/cm²	DOC delivered to the membranes

V_i	L	Initial feed volume
V_f	L	Remaining feed volume
V_{cp}	L	Volume of concentrate and permeate
C_{DOCi}	mg/L	Initial DOC concentration of feed
C_{DOCf}	mg/L	Final DOC concentration of feed
C_{DOCcp}	mg/L	Final DOC concentration of concentrate and permeate
Q^2		Leave-one-out goodness of prediction

Abbreviation	Description
AAE	Average absolute error
ANN	Artificial neural network
ANOVA	Analysis of variance
AOP	Advanced oxidation process
ATR-FTIR	Attenuated total reflection Fourier transform infrared
CA	Contact angle
CP	Concentration polarization
DBPs	Disinfectant by products
DOC	Dissolved organic matter
EDCs	Endocrine disrupting compounds
ESI	Electrospray ionisation
FDA	Food and Drug Administration
GAC	Granular activated carbon
HPLC	High performance liquid chromatography
LC-OCD	Liquid chromatography organic carbon detection
LPRO	Low pressure reverse osmosis
MAPE	Mean absolute percent error
MaxAPE	Maximum absolute percent error
MBR	Membrane bioreactor
MLR	Multiple linear regression
MOPAC	Molecular Orbital PACkage
MS	Mass spectrometry
MW	Molecular weight
MWCO	Molecular weight cut-off
MWCO	Molecular weight cut-off
NF	Nanofiltration
NOM	Natural organic matter

PAC	Powder activated carbon
PCA	Principal component analysis
PCPs	Personal care products
PCR	Principal component regression
PhACs	Pharmaceutically active compounds
PLS	Partial least squares
PM3	Parameterized Model number 3
PRESS	Predictive error sum of squares
PTFE	Polytetrafluoroethylene
PWP	Pure water permeability
QSAR	Quantitative structure-activity relationship
RO	Reverse osmosis
SR	Salt rejection
STDE	Standard deviation of error
TSS	Total sum of squares
ULPRO	Ultra-low pressure reverse osmosis
WTP	Drinking water treatment plant
WWTP	Wastewater treatment plant
ZP	Zeta potential

Appendices

Appendix A, Principal Component Analysis

Principal components are linear combinations of the p random variables x_1, x_2, ..., x_p. Geometrically, these linear combinations represent the selection of a new coordinate system obtained by rotating the original system with x_1, x_2, ..., x_p as the coordinate axes. The new axes represent the directions with maximum variability and provide a simpler and more limited description of the covariance or correlation structure. Principal components depend on the covariance Σ or the correlation matrix $\mathbf{R}$ of x_1, x_2, ..., x_p.

Let the random vector $X = [x_1, x_2, ..., x_p]$ have the covariance matrix Σ with eigenvalues $\lambda_1 \geq \lambda_2 \geq ... \geq \lambda_p \geq 0$. The definition of eigenvalues is as follows, for any n × n square matrix $\mathbf{M}$, a number λ and a non-zero vector $\mathbf{e}$ exist that verify:

$$\mathbf{M}\mathbf{e} = \lambda\mathbf{e} \tag{1}$$

λ is called an eigenvalue for $\mathbf{M}$, and $\mathbf{e}$ is called the corresponding eigenvector. In this case Σ is a square matrix, considering the linear combinations:

$$y_1 = \mathbf{a}'_1\mathbf{X} = a_{11}x_1 + a_{12}x_2 + \cdots + a_{1p}x_p$$
$$y_2 = \mathbf{a}'_2\mathbf{X} = a_{21}x_1 + a_{22}x_2 + \cdots + a_{2p}x_p$$
$$\vdots$$
$$y_p = \mathbf{a}'_p\mathbf{X} = a_{p1}x_1 + a_{p2}x_2 + \cdots + a_{pp}x_p \tag{2}$$

And considering

$$\text{Var}(y_i) = \mathbf{a}'_i\Sigma\mathbf{a}_i \qquad i = 1, 2, ..., p \tag{3}$$
$$\text{Cov}(y_i, y_k) = \mathbf{a}'_i\Sigma\mathbf{a}_k \qquad i, k = 1, 2, ..., p \tag{4}$$

The principal components are those uncorrelated linear combinations y_1, y_2, ..., y_p of the original variables x_1, x_2, ..., x_p whose variances in Eq. 2.4 are as large as possible. They account for maximal proportions of the variation in the original data, i.e. y_1 accounts for the maximum amount of the variance among all possible linear combinations of x_1, ..., x_p, y_2 accounts for the maximum variance subject to being uncorrelated with y_1 and so on. The first principal component (PC) is the linear combination (lc) with maximum variance; that is, maximizes Var (y_1) = $\mathbf{a}'_1\mathbf{\Sigma}\mathbf{a}_1$. It is clear that Var (y_1) = $\mathbf{a}'_1\mathbf{\Sigma}\mathbf{a}_1$ can be increased by multiplying any $\mathbf{a}_1$ by some constant. To eliminate this indeterminacy, it is convenient to restrict attention to coefficient vectors of unit length. This applies since the variances of the principal component variables could be increased without limit, simply by increasing the coefficients that define them; the restriction applied is that the sum of squares of the coefficients is one, in that way the total variance of all the components is equal to the total variance of all the observed variables. Therefore the definitions are:

1st PC = lc $\mathbf{a}'_1\mathbf{X}$ that maximizes Var $(\mathbf{a}'_1\mathbf{X})$ subject to $\mathbf{a}'_1\mathbf{a}_1 = 1$
2nd PC = lc $\mathbf{a}'_2\mathbf{X}$ that maximizes Var $(\mathbf{a}'_2\mathbf{X})$ subject to $\mathbf{a}'_2\mathbf{a}_2 = 1$ and Cov $(\mathbf{a}'_1\mathbf{X}, \mathbf{a}'_2\mathbf{X}) = 0$

At the ith component,

i^{th} PC = lc $\mathbf{a}'_i\mathbf{X}$ that maximizes Var $(\mathbf{a}'_i\mathbf{X})$ subject to $\mathbf{a}'_i\mathbf{a}_i = 1$ and Cov $(\mathbf{a}'_i\mathbf{X}, \mathbf{a}'_k\mathbf{X}) = 0$ for $k < i$

It is often convenient to rescale the coefficients so that their sum of squares is equal to the variance of the component they define. In the case of components derived from the correlation matrix $\mathbf{R}$, these rescaled coefficients give the correlations between the components and the original variables. Those values are often presented as the result of a principal component analysis. The coefficients defining the principal components are given by what are known as the eigenvectors of the covariance $\mathbf{\Sigma}$ or the correlation matrix, $\mathbf{R}$. The full set of principal components is as large as the original set of variables. However in practice, what we see is that the sum of the variances of the first few principal components may exceed 80% of the total variance of the original data. By examining these few new variables, we are able to develop a deeper understanding of the relationships that generated the original data.

Factor analysis can be considered an extension of PCA. Both are attempts to approximate the covariance, $\mathbf{\Sigma}$, or correlation matrix, $\mathbf{R}$ (Landau and Everitt, 2004). From now on, the term "component" is used instead of "factor" for the sake of understanding. Component analysis has the purpose of describing of the correlation or covariance relationships among many variables in terms of a few underlying but unobservable random quantities named components. The argument that motivates a component model is that variables can be grouped by their correlations. That is, when variables within a particular group are highly correlated among themselves, but have relatively small correlations with variables in a different group, then it is conceivable that each group of variables represents a single underlying construct or component that is responsible for the observed correlations. In general terms, component analysis is concerned with whether the covariances or correlations between a set of observed variables x_1, x_2, ..., x_p can be explained in terms of a smaller number of unobservable latent variables or common components, f_1, f_2, ..., f_k, where $k < p$. The formal model linking measured and latent variables is that of multiple regression, with each observed variable being regressed on the common components. The coefficients in the model are known in this context as the *component loadings* (**f**), and the random error terms as *errors* (**u**) since they now represent that part of an observed variable not accounted for by the common components. The orthogonal component model can be written as follows:

$$x_1 = \lambda_{11}f_1 + \lambda_{12}f_2 + \cdots + \lambda_{1k}f_k + u_1$$
$$x_2 = \lambda_{21}f_1 + \lambda_{22}f_2 + \cdots + \lambda_{2k}f_k + u_2$$
$$\vdots$$
$$x_p = \lambda_{p1}f_1 + \lambda_{p2}f_2 + \cdots + \lambda_{pk}f_k + u_p$$

(5)

In vectorial form equations above can be written as

$$\mathbf{x} = \mathbf{\Lambda f} + \mathbf{u} \tag{6}$$

where,

$$\mathbf{x} = \begin{bmatrix} x_1 \\ \vdots \\ x_p \end{bmatrix}, \quad \mathbf{\Lambda} = \begin{bmatrix} \lambda_{11} \cdots \lambda_{1k} \\ \vdots \quad \vdots \\ \lambda_{p1} \cdots \lambda_{pk} \end{bmatrix}, \quad \mathbf{f} = \begin{bmatrix} f_1 \\ \vdots \\ f_k \end{bmatrix}, \quad \mathbf{u} = \begin{bmatrix} u_1 \\ \vdots \\ u_p \end{bmatrix}$$

The location and scale of the unobserved components is assumed, but a valid assumption is a standardised form with a mean of zero and a standard deviation of one. Another assumption is that the error terms are uncorrelated with each other and with the common components. This implies that, given the values of the components, the measured variables are independent, thus the correlations of the observed variables arise from their relationships with the components. Since the components are unobserved, the component loadings cannot be estimated in the same way as are regression coefficients in multiple regressions, in which the independent variables can be observed. But with the assumptions above, the component model implies that the population covariance matrix of the observed variables Σ has the form

$$\Sigma = \mathbf{\Lambda \Lambda}^{\mathrm{T}} + \mathbf{\Psi} \tag{7}$$

where $\mathbf{\Psi}$ is a diagonal matrix containing variances of the error terms on its main diagonal, and this relationship can be used as the basis for estimating both the component loadings and the specific variances. The main methods of estimation for Eq. 7 are the principal component method and the maximum likelihood method. The theory and details of both are given in an advanced book (Johnson and Wichern, 2007). The first method operates much like principal component analysis but only tries to account for the common component variance, and the second relies on assuming that the observed variables have a multivariate normal distribution. All component loadings obtained from the initial loadings by an orthogonal transformation have the same ability to reproduce the covariance or correlation matrix. However, an orthogonal transformation corresponds to a rigid rotation of the coordinate axes. For this reason, an orthogonal transformation of the component loadings, as well as the implied orthogonal transformation of the components, is called component rotation. Thus, since the original loadings may not be readily interpretable, it is usual practice to rotate them until a simpler structure is achieved. Ideally, what is intended is to see a pattern of loadings such that each variable loads high on a single component and has small to moderate loadings of the remaining components (Johnson and Wichern, 2007). Rotation does not alter the overall structure of a solution, but only how the solution is described; the rotation of components is a process by which a solution is made more interpretable without changing its underlying mathematical properties.

Appendix B, Artificial Neural Networks

Definition

An artificial neural network consists of a set of neurons which communicates by sending signals to each other over a large number of weighted connections (Kröse and Van der Smagt, 1996). Artificial neural networks (ANNs) are computational techniques used for modelling linear or non-linear relationships between a number of inputs and outputs. According to Kröse and Van der Smagt, many of these properties can be attributed to existing (non-neural) models; the question is to which extent the neural approach proves to be better suited for certain applications than existing models. The answer is that ANNs have the advantage of handling information with the ability to learn, to generalise, or to cluster or organise data.

Neuron model and transfer functions

A neuron with a scalar input and no bias appears in Fig. 1a. The scalar input p is multiplied by the scalar weight w to form the product wp. The weighted input wp is the argument of the transfer function f, which produces the scalar output a. The neuron has a scalar bias, b. The bias is added to the product wp to shift the function f by an amount b. The bias is much like a weight, except that it has a constant input of 1 (Hagan et al., 2002; Demuth et al., 2007).

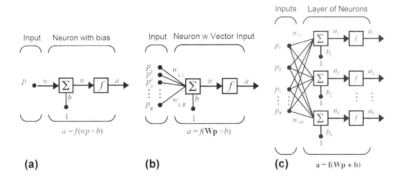

Fig. 1: Input and neuron configuration, a) neuron with scalar input, b) neuron with vector input, c) layer of neurons

Adapted from Demuth et al., 2007

182

The transfer function, net input n, is the sum of the weighted input wp and the bias b. This sum is the argument of the transfer function f. The transfer function can be a step function, a linear function or a sigmoid function (Fig. 2) that takes the argument n and produces the output a. The weight or bias parameters can be adjusted.

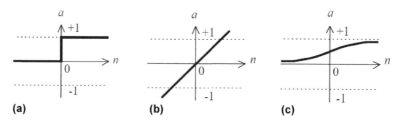

Fig. 2: Transfer functions, a) step function, b) linear function, c) sigmoid function

Adapted from Demuth et al., 2007

A neuron can receive input of an input vector $\mathbf{p}$ (p_1, p_2, ... p_R) as shown in Fig. 1b. Then the individual elements of vector $\mathbf{p}$ are multiplied by weights of the weight vector $\mathbf{W}$. Their sum is $\mathbf{Wp}$, the dot product of the (single row) matrix $\mathbf{W}$ and the vector $\mathbf{p}$. The neuron has a bias $\mathbf{b}$ which is summed with the weighted inputs to form the net input $\mathbf{n}$. This sum, $\mathbf{n}$, is the argument of the transfer function $\mathbf{f}$. Two or more neurons can be combined in a layer, and a particular network could contain one or more such layers. A one-layer network with R input elements and S neurons is shown in Fig. 1c. In this network, each element of the input vector $\mathbf{p}$ is connected to each neuron input through the weight matrix $\mathbf{W}$. The ith neuron has a sum that gathers its weighted inputs and bias to form its own scalar output $n(i)$. The various $n(i)$ taken together form an S-element net input vector $\mathbf{n}$. Finally, the neuron layer outputs form a column vector $\mathbf{a}$.

Neural network architecture

A network with three layers is shown in Fig. 3. Each layer has its own weight matrix $\mathbf{W}$, its own bias vector $\mathbf{b}$, a net input vector $\mathbf{n}$ and an output vector $\mathbf{a}$. The vectors $\mathbf{p}$ and $\mathbf{W}$ can be drawn in abbreviated notation for each layer. The R denotes the length of vector $\mathbf{p}$, and $\mathbf{a}$ and $\mathbf{b}$ are vectors of length S. As defined previously, each layer includes the weight matrix, the summation and multiplication operations, the bias vector $\mathbf{b}$, the transfer function boxes and the output vector. Superscripts are used to identify the layers. A layer whose output is the network output is called an output layer. The other layers are called hidden layers. The network shown in Fig. 3 has an output layer (layer 3) and two hidden layers, layers 1 and 2 (Demuth et al., 2007).

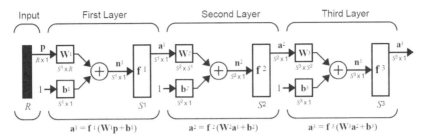

Fig. 3: Neural network with layers

(Hagan et al., 2002)

The ANN architecture used in this thesis is a feedforward network. In feedforward networks the dataflow from input to output neurons is strictly feedforward. The data processing can extend over multiple layers of neurons, but no feedback connections are present. A feed-forward network has a layered structure. Each layer consists of neurons which receive their input from neurons from a preceding layer and send their output to neurons in a subsequent layer. There are no connections within a layer.

Feedforward networks often have one or more hidden layers of sigmoid neurons followed by an output layer of linear or sigmoid neurons. Multiple layers of neurons with nonlinear transfer functions allow the network to learn nonlinear and linear relationships between input and output vectors. The linear output layer allows the network to produce values outside the range −1 to +1. For a range between 0 and 1, the output layer should use a sigmoid transfer function. A two-layer network having a sigmoid first layer and a linear or sigmoid second layer can be trained to approximate most functions (Demuth et al., 2007). This architecture is shown in Fig. 4.

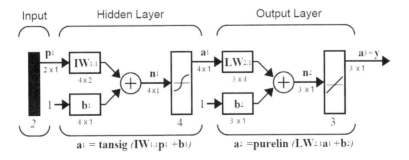

Fig. 4: Feedforward neural network

(Demuth et al., 2007)

Learning process

A neural network has to be configured such that the application of a set of neurons produces the desired set of outputs. Various methods to set the strengths of the connections exist. One way is to set the weights explicitly, using *a priori* knowledge. Another way is to 'train' the neural network by feeding it teaching patterns and letting it change its weights according to some learning rule (Kröse and Van der Smagt, 1996). One method of the learning rule is the back-propagation learning rule, in which the errors for the neurons of the hidden layer are determined by back-propagating the errors of the neurons of the output layer.

During the learning phase, representative examples obtained by experimental sets are presented to the network so that it can integrate this knowledge within its structure. The learning process consists of determining the weights that are produced from the inputs the best fit of the predicted outputs over the entire training data set. The difference between the computed output vector and the target vector is used to determine the weights using an optimization procedure in order to minimise the sum of squares of the errors. The errors between network outputs and targets are summed over the entire training data set and the weights are updated after every presentation of the complete data set (Delgrange et al., 1998).

The learning phase uses two separate datasets, one is the training dataset and the other is the validation dataset. After defining an adequate error measure, the neural network training algorithms try to minimise the error of the set of learning samples and update the weights. The validation dataset is used as the network error test; the training is carried out as long as the training reduces the network's error on the validation dataset. During validation the error begins to drop, then reaches a minimum and finally increases. Continuing the learning process after the point when the validation error arrives at a minimum leads to a process called overfitting. The use of a validation dataset and the fact that training is stopped after the reduction of error of the validation dataset avoids the problem of overfitting. After finishing the learning process (training and validation), another dataset called testing or prediction dataset is used to finally test and confirm the prediction accuracy (Delgrange et al., 1998; Demuth et al., 2007).

Appendix C, Concentration polarization in NF membrane elements

Assumptions:

Small membrane strips (in envelope configuration) that are simulating pieces of flat-sheet membrane.

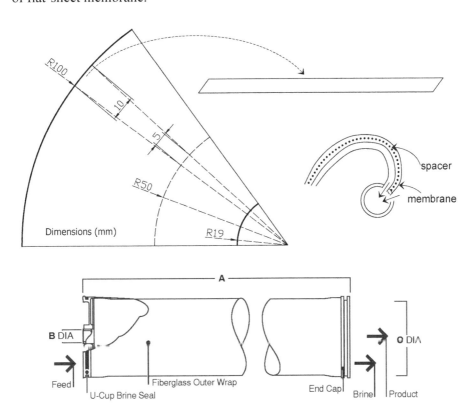

Product specifications

Product	Nominal Active Surface Area ft² (m²)	Single-element Max. recovery	Dimensions – Inches (mm)		
			A	B	C
NF90-400	400 (37)	15%	40 (1,016)	1.5 (38)	7.9 (201)
NF200-400	400 (37)	15%	40 (1,016)	1.5 (38)	7.9 (201)

Variations in concentrate and permeate concentrations were calculated using formulae derived by Mulder (1996):

$$c_r = c_f (1-S)^{-R}$$

$$\bar{c}_p = \frac{c_f}{S}[1-(1-S)^{1-R}]$$

where c_r is concentrate concentration, c_f is the feed concentration, $\bar{c}_p$ is the mean permeate concentration, S is recovery, and R is rejection.

A high flux of 24L/m²-h was assumed in order to obtain a high concentration polarization, lower fluxes will result in lower CP.

Concentration polarisation (or beta factor) for sodium chloride is calculated with the software IMS Design (Hydranautics) by

$$CP = K_p \exp(\frac{2S}{2-S})$$

where K_p is a proportionality constant depending on system geometry, and S is recovery. The equation assumes an average feed flow, that results from the arithmetic average of feed and concentrate flow. For a typical spiral wound element, the K_p is 1.11. The following table shows calculations of CP at different recoveries.

S	0.08	0.15	0.18
CP	1.2	1.3	1.4

The formula does not take into account variation of concetrations of feed and concentrate, and it only applies for sodium chloride. In addition, the system geometry is assumed constant and does not consider changes of CP in the element.

Concentration polarization for clean NF-200

Parameter	Unit	NF-200			
Membrane area (A_m)	cm²	100	200	200	200
Width (cm)	cm	0.5	1.0	1.0	1.0
Cross-section area (A_{cross})	cm²	0.02	0.02	0.02	0.02
Cross flow velocity (U)	cm/s	38.3	76.7	37.8	37.8
Mean diffusion coeff. (D)	cm²/s	6.30E-06	6.30E-06	6.30E-06	6.30E-06
Equiv. hydraulic diameter	cm	0.08	0.08	0.08	0.08
Channel length (L)	cm	100	100	100	100
Mean back diffusion mass transf. coef. (k)	cm/s	1.0E-03	1.3E-03	1.0E-03	1.0E-03
Permeate flow (Q_p)	mL/min	4.0	8.0	8.0	8.0
Concentrate flow, (Q_c)	mL/min	46	92	45	45
Flux, $J = Q_p/A_m$	L/m²-h	24.0	24.0	24.0	24.0
	cm/s	6.7E-04	6.7E-04	6.7E-04	6.7E-04
J/k		0.7	0.5	0.7	0.7
Recovery	%	8	8	15*	15[†]
CP average org. sol.		1.6	1.4	1.6	1.7
c_m average org. sol.	µg/L	12.2	11.1	12.8	21.2
$E_0 = c_p/c_m$		0.254	0.280	0.241	0.189
k (NaCl)		2.0E-03	2.5E-03	2.0E-03	2.0E-03
k (MgSO₄)		1.5E-03	1.9E-03	1.5E-03	1.5E-03
CP (NaCl)		1.3	1.2	1.3	1.3
c_m (NaCl)	µg/L	2.7E+06	2.5E+06	2.8E+06	5.0E+06
$E_0 = c_p/c_m$ (NaCl)		0.246	0.259	0.242	0.183
CP (MgSO₄)		1.5	1.4	1.5	1.6
c_m (MgSO₄)	µg/L	3.4E+06	3.1E+06	3.6E+06	9.8E+06
$E_0 = c_p/c_m$ (MgSO₄)		0.017	0.018	0.016	0.009

* Max. recovery per element.
[†] Total recovery of 70%.

188

Concentration polarization for clean NF-90

Parameter	Unit	NF-90			
Membrane area (A_m)	cm²	100	200	200	200
Width (cm)	cm	0.5	1.0	1.0	1.0
Cross-section area (A_{cross})	cm²	0.02	0.02	0.02	0.02
Cross flow velocity (U)	cm/s	38.3	76.7	37.8	37.8
Mean diffusion coeff. (D)	cm²/s	6.30E-06	6.30E-06	6.30E-06	6.30E-06
Equiv. hydraulic diameter	cm	0.08	0.08	0.08	0.08
Channel length (L)	cm	100	100	100	100
Mean back diffusion mass transf. coef. (k)	cm/s	1.0E-03	1.3E-03	1.0E-03	1.0E-03
Permeate flow (Q_p)	mL/min	4.0	8.0	8.0	8.0
Concentrate flow, (Q_c)	mL/min	46	92	45	45
Flux, $J = Q_p/A_m$	L/m²-h	24.0	24.0	24.0	24.0
	cm/s	6.7E-04	6.7E-04	6.7E-04	6.7E-04
J/k		0.7	0.5	0.7	0.7
Recovery	%	8	8	15*	15[†]
CP average org. sol.		1.9	1.6	1.9	1.9
c_m average org. sol.	µg/L	15.6	13.7	16.9	40.5
$E_0 = c_p/c_m$		0.051	0.058	0.053	0.032
k (NaCl)		2.0E-03	2.5E-03	2.0E-03	2.0E-03
k (MgSO₄)		1.5E-03	1.9E-03	1.5E-03	1.5E-03
CP (NaCl)		1.4	1.3	1.4	1.4
c_m (NaCl)	µg/L	3.0E+06	2.8E+06	3.1E+06	7.4E+06
$E_0 = c_p/c_m$ (NaCl)		0.077	0.082	0.077	0.048
CP (MgSO₄)		1.5	1.4	1.6	1.6
c_m (MgSO₄)	µg/L	3.4E+06	3.1E+06	3.6E+06	9.9E+06
$E_0 = c_p/c_m$ (MgSO₄)		0.010	0.011	0.010	0.006

* Max. recovery per element.
† Total recovery of 70%.

Appendix D, Internal dataset of rejections

Internal experimental rejection database, continued on next page

#	Name	log D	length	depth	eqwidth	membrane	MWCO	SR	Meas. reject.
1	Acetaminophen	0.23	1.14	0.42	0.53	NF90C	200	0.98	71.2
2	Acetaminophen	0.23	1.14	0.42	0.53	NF90C	200	0.98	80.3
3	Acetaminophen	0.23	1.14	0.42	0.53	NF90C	200	0.98	62.4
4	Acetaminophen	0.23	1.14	0.42	0.53	NF200F	300	0.96	17.7
5	Acetaminophen	0.23	1.14	0.42	0.53	NF90F	200	0.98	81.0
6	Phenacetine	1.68	1.35	0.42	0.54	NF200C	300	0.96	41.3
7	Phenacetine	1.68	1.35	0.42	0.54	NF200C	300	0.96	69.6
8	Phenacetine	1.68	1.35	0.42	0.54	NF90C	200	0.98	75.0
9	Phenacetine	1.68	1.35	0.42	0.54	NF90C	200	0.98	77.6
10	Phenacetine	1.68	1.35	0.42	0.54	NF90C	200	0.98	70.9
11	Phenacetine	1.68	1.35	0.42	0.54	NF200F	300	0.96	21.4
12	Phenacetine	1.68	1.35	0.42	0.54	NF90F	200	0.98	76.0
13	Caffeine	-0.45	0.98	0.56	0.70	NF200C	300	0.96	50.0
14	Caffeine	-0.45	0.98	0.56	0.70	NF200C	300	0.96	50.0
15	Caffeine	-0.45	0.98	0.56	0.70	NF200C	300	0.96	50.0
16	Caffeine	-0.45	0.98	0.56	0.70	NF90C	200	0.98	80.8
17	Caffeine	-0.45	0.98	0.56	0.70	NF90C	200	0.98	93.4
18	Caffeine	-0.45	0.98	0.56	0.70	NF90C	200	0.98	80.8
19	Caffeine	-0.45	0.98	0.56	0.70	NF200F	300	0.96	61.9
20	Caffeine	-0.45	0.98	0.56	0.70	NF90F	200	0.98	91.0
21	Metronidazole	-0.27	0.93	0.48	0.66	NF200C	300	0.96	47.3
22	Metronidazole	-0.27	0.93	0.48	0.66	NF200C	300	0.96	41.8
23	Metronidazole	-0.27	0.93	0.48	0.66	NF200C	300	0.96	34.5
24	Metronidazole	-0.27	0.93	0.48	0.66	NF90C	200	0.98	82.5
25	Metronidazole	-0.27	0.93	0.48	0.66	NF90C	200	0.98	92.8
26	Metronidazole	-0.27	0.93	0.48	0.66	NF90C	200	0.98	88.3
27	Metronidazole	-0.27	0.93	0.48	0.66	NF200F	300	0.96	27.6
28	Metronidazole	-0.27	0.93	0.48	0.66	NF90F	200	0.98	90.1
29	Phenazone	0.54	1.17	0.56	0.66	NF200C	300	0.96	52.5
30	Phenazone	0.54	1.17	0.56	0.66	NF200C	300	0.96	69.2
31	Phenazone	0.54	1.17	0.56	0.66	NF200C	300	0.96	61.7
32	Phenazone	0.54	1.17	0.56	0.66	NF90C	200	0.98	85.0
33	Phenazone	0.54	1.17	0.56	0.66	NF90C	200	0.98	95.8
34	Phenazone	0.54	1.17	0.56	0.66	NF90C	200	0.98	95.6
35	Phenazone	0.54	1.17	0.56	0.66	NF200F	300	0.96	56.4
36	Phenazone	0.54	1.17	0.56	0.66	NF90F	200	0.98	93.9

Internal experimental rejection database, continued on next page

#	Name	log D	length	depth	eqwidth	membrane	MWCO	SR	Meas. reject.
37	Sulphamethoxazole	-0.45	1.33	0.58	0.64	NF200C	300	0.96	58.9
38	Sulphamethoxazole	-0.45	1.33	0.58	0.64	NF200C	300	0.96	66.1
39	Sulphamethoxazole	-0.45	1.33	0.58	0.64	NF200C	300	0.96	71.4
40	Sulphamethoxazole	-0.45	1.33	0.58	0.64	NF90C	200	0.98	94.4
41	Sulphamethoxazole	-0.45	1.33	0.58	0.64	NF90C	200	0.98	98.3
42	Sulphamethoxazole	-0.45	1.33	0.58	0.64	NF90C	200	0.98	98.5
43	Sulphamethoxazole	-0.45	1.33	0.58	0.64	NF200F	300	0.96	48.8
44	Sulphamethoxazole	-0.45	1.33	0.58	0.64	NF90F	200	0.98	96.5
45	Carbamazepine	2.58	1.20	0.58	0.73	NF200C	300	0.96	70.0
46	Carbamazepine	2.58	1.20	0.58	0.73	NF200C	300	0.96	74.3
47	Carbamazepine	2.58	1.20	0.58	0.73	NF200C	300	0.96	72.9
48	Carbamazepine	2.58	1.20	0.58	0.73	NF90C	200	0.98	90.8
49	Carbamazepine	2.58	1.20	0.58	0.73	NF90C	200	0.98	97.3
50	Carbamazepine	2.58	1.20	0.58	0.73	NF90C	200	0.98	98.2
51	Carbamazepine	2.58	1.20	0.58	0.73	NF200F	300	0.96	73.0
52	Carbamazepine	2.58	1.20	0.58	0.73	NF90F	200	0.98	94.5
53	Atrazine	2.52	1.26	0.55	0.74	NF200C	300	0.96	81.3
54	Atrazine	2.52	1.26	0.55	0.74	NF200C	300	0.96	82.5
55	Atrazine	2.52	1.26	0.55	0.74	NF200C	300	0.96	83.8
56	Atrazine	2.52	1.26	0.55	0.74	NF90C	200	0.98	95.0
57	Atrazine	2.52	1.26	0.55	0.74	NF90C	200	0.98	97.8
58	Atrazine	2.52	1.26	0.55	0.74	NF90C	200	0.98	97.8
59	Atrazine	2.52	1.26	0.55	0.74	NF200F	300	0.96	88.0
60	Atrazine	2.52	1.26	0.55	0.74	NF90F	200	0.98	97.0
61	Naproxen	0.34	1.37	0.75	0.76	NF200C	300	0.96	75.6
62	Naproxen	0.34	1.37	0.75	0.76	NF200C	300	0.96	91.5
63	Naproxen	0.34	1.37	0.75	0.76	NF200C	300	0.96	93.9
64	Naproxen	0.34	1.37	0.75	0.76	NF90C	200	0.98	96.0
65	Naproxen	0.34	1.37	0.75	0.76	NF90C	200	0.98	98.9
66	Naproxen	0.34	1.37	0.75	0.76	NF90C	200	0.98	99.0
67	Naproxen	0.34	1.37	0.75	0.76	NF200F	300	0.96	79.7
68	Naproxen	0.34	1.37	0.75	0.76	NF90F	200	0.98	96.5
69	Ibuprofen	0.77	1.39	0.55	0.64	NF200C	300	0.96	75.5
70	Ibuprofen	0.77	1.39	0.55	0.64	NF200C	300	0.96	89.1
71	Ibuprofen	0.77	1.39	0.55	0.64	NF200C	300	0.96	93.8
72	Ibuprofen	0.77	1.39	0.55	0.64	NF90C	200	0.98	96.0
73	Ibuprofen	0.77	1.39	0.55	0.64	NF90C	200	0.98	98.8
74	Ibuprofen	0.77	1.39	0.55	0.64	NF90C	200	0.98	99.0
75	Ibuprofen	0.77	1.39	0.55	0.64	NF200F	300	0.96	87.5
76	Ibuprofen	0.77	1.39	0.55	0.64	NF90F	200	0.98	97.0

Internal experimental rejection database

#	Name	log D	length	depth	eqwidth	membrane	MWCO	SR	Meas. reject.
77	17beta estradiol	3.94	1.39	0.65	0.74	NF200C	300	0.96	63.2
78	17beta estradiol	3.94	1.39	0.65	0.74	NF200C	300	0.96	75.8
79	17beta estradiol	3.94	1.39	0.65	0.74	NF200C	300	0.96	60.5
80	17beta estradiol	3.94	1.39	0.65	0.74	NF90C	200	0.98	90.9
81	17beta estradiol	3.94	1.39	0.65	0.74	NF90C	200	0.98	97.8
82	17beta estradiol	3.94	1.39	0.65	0.74	NF90C	200	0.98	95.3
83	17beta estradiol	3.94	1.39	0.65	0.74	NF200F	300	0.96	76.5
84	17beta estradiol	3.94	1.39	0.65	0.74	NF90F	200	0.98	97.8
85	Estrone	3.46	1.39	0.67	0.76	NF200C	300	0.96	76.4
86	Estrone	3.46	1.39	0.67	0.76	NF200C	300	0.96	77.3
87	Estrone	3.46	1.39	0.67	0.76	NF200C	300	0.96	57.3
88	Estrone	3.46	1.39	0.67	0.76	NF90C	200	0.98	90.3
89	Estrone	3.46	1.39	0.67	0.76	NF90C	200	0.98	97.5
90	Estrone	3.46	1.39	0.67	0.76	NF90C	200	0.98	92.2
91	Estrone	3.46	1.39	0.67	0.76	NF200F	300	0.96	79.2
92	Estrone	3.46	1.39	0.67	0.76	NF90F	200	0.98	96.9
93	Nonylphenol	5.88	1.79	0.59	0.66	NF200C	300	0.96	83.3
94	Nonylphenol	5.88	1.79	0.59	0.66	NF200C	300	0.96	83.3
95	Nonylphenol	5.88	1.79	0.59	0.66	NF200C	300	0.96	83.3
96	Nonylphenol	5.88	1.79	0.59	0.66	NF90C	200	0.98	90.3
97	Nonylphenol	5.88	1.79	0.59	0.66	NF90C	200	0.98	97.7
98	Nonylphenol	5.88	1.79	0.59	0.66	NF90C	200	0.98	97.8
99	Nonylphenol	5.88	1.79	0.59	0.66	NF200F	200	0.96	92.7
100	Nonylphenol	5.88	1.79	0.59	0.66	NF90F	200	0.98	97.6
101	Bisphenol A	3.86	1.25	0.75	0.79	NF200C	300	0.96	45.4
102	Bisphenol A	3.86	1.25	0.75	0.79	NF90C	200	0.98	90.4
103	Bisphenol A	3.86	1.25	0.75	0.79	NF90C	200	0.98	97.1
104	Bisphenol A	3.86	1.25	0.75	0.79	NF90C	200	0.98	95.0
105	Bisphenol A	3.86	1.25	0.75	0.79	NF200F	300	0.96	51.0
106	Bisphenol A	3.86	1.25	0.75	0.79	NF90F	200	0.98	94.6

192

Appendix E, External dataset of rejections

NF-90

Compound	log D	length	depth	eqwidth	SR	Measured rejection (%)
Dichloroacetic acid	-3.54	0.7	0.52	0.6	0.99	89
Trichloroacetic acid	-2.42	0.89	0.41	0.52	0.99	87
Ibuprofen	1.15	1.31	0.64	0.7	0.99	86
Diclofenac	0.48	1.13	0.45	0.64	0.99	90
Clofibric acid	-0.91	0.95	0.41	0.52	0.99	86
Naproxen	0.41	1.26	0.54	0.61	0.99	89
Chloroform	1.97	0.53	0.35	0.42	0.99	0
Primidone	-0.84	0.97	0.48	0.65	0.99	82
Perchloroethene	3.4	0.78	0.45	0.59	0.99	39
Carbontetrachloride	2.83	0.64	0.57	0.6	0.99	35

Trisep TS-80

Compound	log D	length	depth	eqwidth	SR	Measured rejection (%)
Atenolol	-2.21	1.78	0.61	0.69	0.98	89
Atropine	-1.22	1.26	0.87	0.92	0.98	95
Clenbuterol	0.26	1.40	0.74	0.81	0.98	87
Dikegulac	-2.05	1.19	0.69	0.81	0.98	95
Metoprolol	-0.77	1.86	0.62	0.73	0.98	90
Metribuzin	0.47	1.17	0.64	0.74	0.98	97
Monomethylphtalate	-2.27	0.99	0.47	0.63	0.98	94
N-acetyl-L-tyrosine	-2.18	1.33	0.60	0.71	0.98	94
Salbutamol	-1.38	1.36	0.74	0.80	0.98	94
Sotalol	-1.70	1.47	0.63	0.78	0.98	90
Terbutaline	-1.64	1.34	0.70	0.77	0.98	89
2-methoxyethanol	-0.77	0.87	0.51	0.52	0.98	32
Aminopyrine	1.00	1.27	0.66	0.73	0.98	97
Antipyrine	0.38	1.17	0.56	0.66	0.98	85
Cyclophosphamide	0.63	1.12	0.63	0.78	0.98	94
Ethanol	-0.31	0.64	0.51	0.52	0.98	9
Glucose	-3.24	0.94	0.71	0.75	0.98	98
Metoxuron	1.64	1.29	0.59	0.70	0.98	89
Pentoxifylline	0.29	1.52	0.70	0.81	0.98	94
Sucrose	-3.70	1.27	0.86	0.91	0.98	99
Clofibric acid	-0.20	1.20	0.68	0.73	0.98	98
Diclofenac	1.62	1.16	0.85	0.90	0.98	99
Fenoprofen	0.37	1.16	0.74	0.83	0.98	99
Gemfibrozil	2.30	1.58	0.65	0.78	0.98	98
Ibuprofen	0.77	1.33	0.70	0.71	0.98	99

Ketoprofen	-0.13	1.16	0.74	0.83	0.98	99
Naproxen	0.34	1.37	0.76	0.76	0.98	95
Atrazine	2.61	1.26	0.55	0.74	0.98	91
Bentazon	2.34	1.18	0.73	0.76	0.98	98
Carbamazepine	2.45	1.20	0.58	0.73	0.98	88
Chlorotoluron	2.41	1.29	0.45	0.61	0.98	81
Diuron	2.68	1.31	0.42	0.56	0.98	73
Estrone	3.13	1.39	0.67	0.76	0.98	90
Isoproturon	2.87	1.42	0.63	0.66	0.98	93
Metobromuron	2.38	1.34	0.50	0.61	0.98	78
Monolinuron	2.30	1.22	0.65	0.69	0.98	79
Simazin	2.18	1.37	0.48	0.64	0.98	82

Desal HL

Compound	log D	length	depth	eqwidth	SR	Measured rejection (%)
Atenolol	-2.21	1.78	0.61	0.69	0.97	95
Atropine	-1.22	1.26	0.87	0.92	0.97	98
Metoprolol	-0.77	1.86	0.62	0.73	0.97	95
Pindolol	-0.32	1.46	0.72	0.76	0.97	83
Terbutaline	-1.64	1.34	0.70	0.77	0.97	93
2-(1H)-Quinoline	1.26	1.00	0.36	0.52	0.97	21
2-ethoxyethanol	-0.32	1.00	0.52	0.53	0.97	38
2-methoxyethanol	-0.77	0.87	0.51	0.52	0.97	25
Aminopyrine	1.00	1.27	0.66	0.73	0.97	98
Antipyrine	0.38	1.17	0.56	0.66	0.97	75
Cyclophosphamide	0.63	1.12	0.63	0.78	0.97	97
Glucose	-3.24	0.94	0.71	0.75	0.97	90
Glycerol	-1.76	0.80	0.53	0.56	0.97	12
NDMA	-0.57	0.70	0.44	0.52	0.97	0
Pentoxifylline	0.29	1.52	0.70	0.81	0.97	99
Sucrose	-3.70	1.27	0.86	0.91	0.97	98
Clofibric acid	-0.20	1.20	0.68	0.73	0.97	99
Diclofenac	1.62	1.16	0.85	0.90	0.97	99
Fenoprofen	0.37	1.16	0.74	0.83	0.97	100
Gemfibrozil	2.30	1.58	0.65	0.78	0.97	99
Ibuprofen	0.77	1.33	0.70	0.71	0.97	99
Ketoprofen	-0.13	1.16	0.74	0.83	0.97	99
Naproxen	0.34	1.37	0.76	0.76	0.97	99
Propranolol	0.55	1.41	0.88	0.90	0.97	87
Carbamazepine	2.45	1.20	0.58	0.73	0.97	95

Appendix F, Correlation matrix

#		1 MW	2 solub.	3 log K$_{ow}$	4 log D	5 dipole	6 MV	7 length	8 width	9 depth	10 eqwidth	11 MWCO	12 PWP	13 SR	14 ZP	15 CA	16 P	17 v	18 k	19 J	20 Jo/k	21 recov	Rejec.
1	MW	1.000																					
2	solubility	-.469	1.000																				
3	log Kow	.541	-.583	1.000																			
4	log D	.476	-.398	.812	1.000																		
5	dipole	-.314	.198	-.706	-.781	1.000																	
6	MV	.782	-.433	.884	.774	-.620	1.000																
7	length	.467	-.441	.865	.707	-.531	.827	1.000															
8	width	.188	-.064	-.048	.161	-.170	-.102	-.383	1.000														
9	depth	.720	-.232	.498	.344	-.499	.631	.312	.180	1.000													
10	eqwidth	.659	-.203	.352	.349	-.458	.440	.029	.665	.851	1.000												
11	MWCO	.060	.000	.023	.002	.005	.040	.020	.052	.032	.054	1.000											
12	PWP	-.055	.000	-.021	-.002	-.004	-.037	-.019	-.048	-.029	-.049	-.982	1.000										
13	SR	-.060	.001	-.023	-.002	-.005	-.040	-.020	-.052	-.032	-.054	-1.000	.982	1.000									
14	ZP	.071	.000	.027	.003	.006	.048	.024	.062	.038	.064	.936	-.930	-.936	1.000								
15	CA	-.062	.001	-.024	-.003	-.005	-.042	-.021	-.054	-.033	-.056	-.991	.990	.991	-.968	1.000							
16	P	-.008	-.009	.003	.014	-.012	.000	-.002	-.011	.009	.001	.297	-.286	-.297	.118	-.244	1.000						
17	v	-.040	.001	-.016	-.003	-.002	-.027	-.014	-.035	-.022	-.037	-.695	.755	.695	-.720	.737	.107	1.000					
18	k	-.044	.002	-.018	-.004	-.002	-.030	-.015	-.038	-.026	-.041	-.793	.843	.793	-.788	.824	-.044	.985	1.000				
19	J	-.052	.002	-.020	-.002	-.005	-.036	-.020	-.044	-.024	-.043	-.942	.947	.942	-.907	.949	-.095	.838	.891	1.000			
20	Jo/k	-.040	-.009	-.009	.013	-.015	-.021	-.013	-.039	-.007	-.027	-.220	.146	.220	-.286	.214	.582	-.059	-.079	.228	1.000		
21	recov	.014	.000	.009	.008	-.067	.010	-.001	.015	.026	.027	.268	-.354	-.268	.271	-.307	.059	-.733	-.708	-.334	.491	1.000	
	rejection	.328	-.194	.413	.239	-.229	.378	.340	.093	.340	.307	-.660	.651	.660	-.579	.643	-.208	.466	.530	.652	.144	-.146	1.000

List of publications

Peer-reviewed scientific journals

Yangali-Quintanilla, V., Kim, T.-U., Kennedy, M. and Amy, G., 2008. Modeling of RO/NF membrane rejections of PhACs and organic compounds: a statistical analysis, Drinking Water Engineering and Science 1 (1), 7-15.

Yangali-Quintanilla, V., Sadmani, A., McConville, M., Kennedy, M. and Amy, G., 2009. Rejection of pharmaceutically active compounds and endocrine disrupting compounds by clean and fouled nanofiltration membranes. Water Research 43 (9), 2349-2362.

Yangali-Quintanilla, V., Verliefde, A., Kim, T.-U., Sadmani, A., Kennedy, M. and Amy, G., 2009. Artificial neural network models based on QSAR for predicting rejection of neutral organic compounds by polyamide nanofiltration and reverse osmosis membranes. Journal of Membrane Science 342 (1-2), 251-262.

Yangali-Quintanilla, V., Sadmani, A., McConville, M., Kennedy, M. and Amy, G., 2009. A QSAR model for predicting rejection of emerging contaminants (pharmaceuticals, endocrine disruptors) by nanofiltration membranes. Water Research, In Press, doi:10.1016/j.watres.2009.06.054.

Yangali-Quintanilla, V., Sadmani, A., Kennedy, M. and Amy, G., 2009. A QSAR (quantitative structure-activity relationship) approach for modelling and prediction of rejection of emerging contaminants by NF membranes. Desalination and Water Treatment, In Press.

Yangali-Quintanilla, V., Maeng, S.K., Fujioka, T., Kennedy, M. and Amy, G., 2009. Nanofiltration as a robust barrier for organic micropollutants in water reuse. Submitted to Environmental Science & Technology.

Kennedy, M., Chun, H., Yangali-Quintanilla, V., Heijman, B. and Schippers, J., 2005. Natural organic matter fouling of UF membranes: fractionation of NOM in surface water and characterisation by LC-OCD, Desalination 178, 73-83.

Conference proceedings

Yangali-Quintanilla, V. Kennedy, M., Heijman, B., Amy, G. and Schippers, J., 2007. Colloidal and non colloidal NOM fouling of UF membranes, AWWA Membrane Technology Conference, Tampa, Florida, USA.

196

Yangali-Quintanilla, V., Verliefde, A., Kennedy, M. and Amy, G., 2007. Enhancement of existing models on transport (rejection) of trace organic compounds through (by) NF/RO membranes. Techneau, Optimal integration of membrane filtration in drinking water treatment, Berlin, Germany.

Yangali-Quintanilla, V., Verliefde, A., Kennedy, M. and Amy, G., 2008. Removal of organic trace pollutants by membrane separation (NF/RO). Techneau, Optimal integration of membrane filtration in drinking water treatment, Trondheim, Norway.

Amy, G., Maeng, S.K., Jekel, M., Ernst, M., Villacorte, L.O., Yangali Quintanilla, V., Kim, T.U. and Reemtsma T., 2008. Advanced water/wastewater treatment process selection for organic micropollutant removal: a quantitative structure-activity relationship (QSAR) approach. In: International Water Week, Water Convention, Singapore.

Yangali-Quintanilla, V., Sadmani, A., McConville, M., Kennedy, M. and Amy, G., 2009. Rejection of Organic Micropollutants by Clean and Fouled Nanofiltration Membranes, AWWA Membrane Technology Conference, Memphis, Tennessee, USA.

Yangali-Quintanilla, Kennedy, M. and Amy, G., 2009. Applications of quantitative structure-activity relationships for rejection of organic solutes by nanofiltration membranes. Techneau, Safe Drinking Water from Source to Tap State-of-the-Art and Perspectives, Maastricht, the Netherlands.

Yangali-Quintanilla, V., Fujioka, T., Kennedy, M. and Amy, G., 2009. Rejection of emerging contaminants by nanofiltration membranes, High Quality Drinking Water Conference, 2009, Delft, the Netherlands.

Yangali-Quintanilla, V., Sadmani, A., Kennedy, M. and Amy, G., 2009. A QSAR (quantitative structure-activity relationship) approach for modelling and prediction of rejection of emerging contaminants by NF and RO membranes. EDS Conference, Desalination for the Environment, Baden-Baden, Germany.

Yangali-Quintanilla, V., Fujioka, T., Kennedy, M. and Amy, G., 2009. Is Nanofiltration a Robust Barrier for Organic Micropollutants. 5th IWA Specialized Membrane Technology Conference for Water and Wastewater Treatment, Beijing, China.

Yangali-Quintanilla, V., Fujioka, T., Kennedy, M. and Amy, G., 2009. Nanofiltration for Removal of PhACs and EDCs, 1st IWA BeNeLux Young Water Professionals Conference, 2009, Eindhoven, the Netherlands.

Book chapters and technical magazines

Yangali-Quintanilla, V., Sadmani, A., Kennedy, M. and Amy, G., 2009. Applications of quantitative structure-activity relationships for rejection modelling of organic solutes by nanofiltration membranes in TECHNEAU: Safe drinking water from source to tap, state of the art & pespectives. Van den Hoven, Theo and Kazner, Christian (eds), IWA Publishing, London, UK.

Verliefde, A. and Yangali-Quintanilla, V., 2008. Removal of pharmaceuticals, Delft Cluster Magazine 5, p.18-21.

Acknowledgements

Acknowledgements to sponsors

Delft Cluster is thanked for its support to UNESCO-IHE and TU Delft for funding this doctoral research as part of project 06.30 Innovatieve drinkwaterzuivering (Innovative drinking water purification).

This doctoral research was also partly sponsored by EU TECHNEAU Project (Work Package 2.3 Optimal Integration of Membrane Filtration in Drinking Water Treatment) under the Sixth Framework Programme (FP6).

The NUFFIC (Netherlands organization for international cooperation in higher education) and KWR (Watercycle Research Institute previously known as Kiwa Water Research) are acknowledged for providing me the opportunity to earn the MSc in Municipal Water and Infrastructure with a specialisation in water treatment before embarking on this PhD research.

Acknowledgements

Recognition of the invaluable help of Prof. Dr. Gary Amy is gratefully acknowledged. His skilful guidance and encouragement helped me to persevere through the 'sometimes' sinusoidal route (with ups and downs) of pursuing a PhD degree. Dr. Maria Kennedy is also gratefully acknowledged for her stimulus and encouragement convincing me to undertake doctoral research. Prof. Amy and Dr. Kennedy are thanked for their support providing help and valuable suggestions towards the execution of this thesis. Their time and knowledge were the inspiration that allowed me to advance through each step of the research.

My collaborators, Julien Chenet, Anwar Sadmani and Takahiro Fujioka, former MSc students at UNESCO-IHE, receive my special gratitude for providing complete assistance during their time at the Institute, while working at the laboratory. All made a good team that, starting from scratch, contributed to reach the pinnacle of this research. Special gratitude is also extended to Megan McConville and Winfrey Humme, who provided assistance by operating instruments and helping with experiments at the laboratory. The generous support of Fred Kruis, Lyzette Robbemont, Peter Heerings, Frank Wiegman and Don van Galen (UNESCO-IHE laboratory staff) is also recognised and was important in all the experimental work.

The support of Dr. Arne Verliefde and Dr. Tae-Uk Kim is acknowledged. Their work was an important cornerstone for me to start performing statistical analyses; then, their results were combined with our results to develop more elaborate models. Thanks to Dr. Gerald Corzo for his introductory lessons on artificial neural networks. I also acknowledge: Dr. Frank Sacher (TZW, Germany) for contributing with the analytical quantification of the organic compounds, Dr. Jaeweon Cho (GIST, Korea) for arranging some measurements of zeta potential, Dr. Arne Verliefde (TU Delft) for contributing measurements of contact angles and zeta potential, Dr. Steven Mookhoek (TU Delft) for arranging equipment for measurement of contact angles, Timothy Selle and Katariina Majamaa (Dow-Filmtec) for kindly providing membrane samples and important information on membranes and software. Sung Kyu Maeng and David de Ridder are acknowledged for contributing with ideas and discussion regarding model validation, and for providing the software MobyDigs (Talete, Milano).

I thank the following people for their help and support: Jolanda Boots, Tanny van der Klis, Maria Laura Sorrentino, Ineke Melis, Fatima Douadi, Ellen de Kok, Karin Priem and Marielle van Erven, all from UNESCO-IHE. I also thank the support of lecturers and colleagues of the Municipal Water & Infrastructure department: Jan-Herman Koster, Jan Peter Buiteman, Saroj Sharma, Kebreab Ghebremichael, Branislav Petrusevski, Carlos Lopez Vasquez, Sergio Salinas Rodriguez, Andrew (Sung Kyu) Maeng, Changwon Ha, Saeed Baghoth, Tarek Waly, Assiyeh Tabatabai, Valentine Uwamariya and Loreen Villacorte.

I am also thankful for the support of the drinking water group at TU Delft (Prof. Hans Van Dijk, Jasper Verberk, Bas Heijman, Arne Verliefde, Luuk Rietveld, Mieke Hubert, Doris Van Halen, Petra Ross, David de Ridder, Sheng Li) and Kiwa Water Research (Emile Cornelissen, Anneke Abrahamse, Sidney Meijering).

The Samenvatting of this thesis was an elegant contribution of Dr. Arne Verliefde, who translated the text to Dutch; and the propositions were translated by Mariska Ronteltap (thanks Mariska).

My wife (Maruja Melva) is gratefully thanked for her great love and for her caring support and encouragement during the long journey of doing a PhD. I thank my family, especially my mother and father, my sister and brothers, and my very large family, for their love and support.

Thanks to my friends for providing many hours of hapiness, conversation and distraction that made more enjoyable this very challenging time of my life.

Curriculum Vitae

Victor Yangali was born in Cobriza, Huancavelica, Peru, in 1976. Cobriza is a mining town of the Andes, exploiting silver and copper; he lived there for ten years, then moved to Huancayo to acquire a secondary education. He received his BSc in Environmental Engineering with a specialisation in sanitary engineering from the National University of Engineering, where he graduated with distinction. Mr. Yangali then received an engineering degree and became part of the engineers association of Peru. He made a short career as a professional engineer in design, construction and supervision of water and wastewater projects. One of his dreams at the university was to pursue studies towards a master of sciences degree in a foreign country, and thus he obtained an MSc degree at UNESCO-IHE, Institute for Water Education.

Education

Mar 1997 – Jul 2000	Environmental Engineering, spec. Sanitary Engineering, National University of Engineering, UNI (Universidad Nacional de Ingeniería, Perú)
Oct 2003 – May 2005	MSc in Municipal Water and Infrastructure. UNESCO-IHE, Institute for Water Education, Netherlands.

Professional experience

01/2008 - 04/2008: Corpei, Corporación Peruana de Ingeniería SA
08/2005 - 01/2006: K-Water (ex Kowaco)
08/2000 - 06/2003: Acruta & Tapia Ingenieros SAC
01/2000 - 07/2000: Sedapal

Memberships

Peruvian Engineers Association (CIP)
European Desalination Society (EDS)
International Desalination Association (IDA)
American Water Works Association (AWWA)
Water Environment Federation (WEF)